AUSTRALIA'S
National Parks

A *Steve Parish* GUIDE

A JOURNEY OF DISCOVERY

To complete your journey, use Steve Parish road atlases and maps

AUSTRALIA'S
NATIONAL PARKS
A *Steve Parish* GUIDE

STATE BY STATE COVERAGE WITH SUPERB PHOTOGRAPHS

Australia's national parks and reserves preserve all that is best about the continent's wilderness areas. They vary in size from enormous to tiny, but I am very conscious that if even the smallest disappeared the nation would be much poorer in natural treasures. Today, awareness of the fragility of wild places is at an all-time high.

The parks included in this book are ones I have particularly enjoyed visiting. They are some of the nation's primary parks, and I have also included in my selection criteria examples of diversity of habitat and human experience. However, there are many other national parks, reserves and conservation areas, and I hope the book inspires its readers to explore as many as possible and to discover their unique qualities.

In creating this book, I have chosen to include as many photographs as possible rather than maps. Detailed, up-to-date maps are best acquired from the park authorities, and their headquarters are listed on page 123. In terms of general access, I suggest that the **Steve Parish Discovery Atlas**, which maps the roads of Australia, makes an ideal companion volume.

Our national parks display the enormous variety and diversity of Australia's natural wonders. Some, mostly in remote, arid or mountainous regions, contain landscapes almost unaltered for many thousands of years. Others have been greatly changed by human activities and present huge challenges to authorities devising restoration programs to restore the parks' integrity.

While taking the photographs in this **National Parks Discovery Guide,** I have watched many people discover our national parks, and their amazement, joy and wonder has never failed to stoke the fires of my own enthusiasm. As you turn the pages, I hope you share my never-ending passion for the landscapes, seascapes and skyscapes of Australia's national parks, and for all the living things which grow, walk, hop, crawl, swim and fly in them.

We'll probably meet out there some day soon.

Steve Parish

CONTENTS

Throughout Australia, areas of land are set aside for recreation, conservation of natural and cultural resources, scientific research, and maintenance of biodiversity, now and for the future. This book explores some of the lesser-known, as well as many of the most popular, national parks and includes some marine parks, nature reserves and World Heritage Areas.

Each of these areas has its own unique character and it may offer a variety of facilities. Each State and Territory is responsible for protected areas within its boundaries; the Commonwealth has responsibility for Booderee and Commonwealth waters; and there are cooperative agreements between Commonwealth and State for Ningaloo and Great Barrier Reef Marine Parks, and between Common-wealth and Aboriginal groups for Kakadu and Uluru-Kata Tjuta. Each publishes information to help plan and enjoy a visit.

It is wisest to find out about conditions for park entry before setting out. For example, an access permit is required to enter or travel over most Aboriginal land (allow plenty of time to get one). For information on the State and Territory bodies to contact regarding access, camping, fees and permits, see page 126.

In this book, facilities available in each park are identified by the pictograms (listed below) on an easily scanned side panel, on which a location map and key features for each park are also shown.

The number of people enjoying Australia's national parks and reserves is ever-increasing. Visitors can avoid placing pressures upon often fragile environments by observing some courtesies:
- Plants, animals and landscape features should be left untouched.
- All rubbish should be removed from the park.
- Pets should not be brought into the park.
- Firewood should not be collected within the park. Campfires should be carefully tended and extinguished. Observe fire bans.
- No pollutants or debris (for example, shampoo, soap or camp rubbish) should be allowed to enter streams or other water.
- Aboriginal and park management restrictions on entry to a park and activities within it should be respected.
- Respect warnings posted by park authorities. They are there to protect the environment and park visitors.
- Take every precaution against carrying plant diseases and noxious weeds into parks. For example, in south-western WA, precautions must be taken to avoid spreading the plant disease "die-back".

KEY TO PICTOGRAMS USED IN THIS BOOK

TOILETS Toilet blocks on site

WALKING TRACKS Established tracks

SNOW SKIING Downhill and/or cross-country skiing

VEHICLE-BASED CAMPING Vehicles can be driven to camp sites

VISITOR CENTRE Park information available

PICNIC AREAS Tables, bins, etc., provided

WHEELCHAIR TRACKS Wheelchair access on established tracks

BIKE TRAILS Bicycles permitted on designated trails

BUSH CAMPING Walk to designated camp sites

SCENIC DRIVES Established scenic road routes

HORSE TRAILS Horses permitted on designated trails

CARAVAN CAMPING Caravan sites available

CABIN ACCOMODATION Motels, huts, home-steads, etc., available

Page 1: Uluru, Uluru-Kata Tjuta NP, NT. Pages 2–3: Ormiston Gorge, West MacDonnell NP, NT. Above: Sunset at Whisky Bay, Wilsons Promontory NP, Victoria

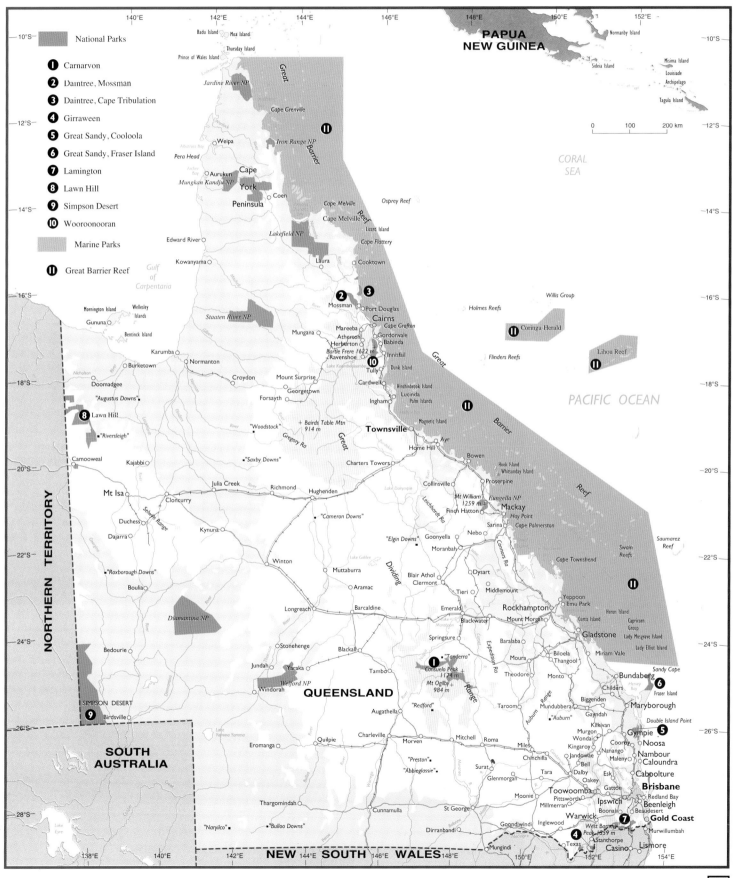

National Parks

1. Carnarvon
2. Daintree, Mossman
3. Daintree, Cape Tribulation
4. Girraween
5. Great Sandy, Cooloola
6. Great Sandy, Fraser Island
7. Lamington
8. Lawn Hill
9. Simpson Desert
10. Wooroonooran

Marine Parks

11. Great Barrier Reef

0   100   200 km

PAPUA NEW GUINEA

CORAL SEA

PACIFIC OCEAN

NORTHERN TERRITORY

SOUTH AUSTRALIA

QUEENSLAND

NEW SOUTH WALES

Gulf of Carpentaria

ACCESS

Binna Burra is 107 km S of Brisbane via Canungra.
Green Mountains is 115 km from Brisbane via Canungra.
Both access roads are narrow and winding and unsuitable for caravans.

CAMPING

Basic facilities at Green Mountains. Bush camping at designated sites February—December. Booking and permit required. Fees charged. Privately-owned accommodation at Green Mountains and Binna Burra.

WHEN TO VISIT

Popular all year round especially during school holidays. Cool days for bushwalking in winter and spring. Winter nights are cold.

SAFETY

Walking tracks slippery in wet weather. Care is needed on tracks edging steep cliffs. Take warm and wet-weather gear in case of sudden storms and temperature drops.

Bracket fungus in the rainforest.

Lamington is one of several national parks on the deeply eroded, volcanic plateau known as the Scenic Rim. It is rugged country with the jagged peaks of the McPherson Range rising to almost 1200 m above sea level.

Mountain creeks plummet from sheer-faced cliffs through deep, narrow valleys. Large tracts of subtropical rainforest grow on the rich volcanic soils. Woody vines, ferns and orchids compete for space and sunlight beneath the canopy of tall trees.

On the slopes above 1000 m in areas of cool temperate rainforest, there are stands of Antarctic beech, some of which may be more than 1000 years old. To walk in the damp mist among these gnarled, moss-covered trees is to experience the feeling of ancient times, when rainforest covered much of Australia, before climate changes took effect.

Above: Elabana Falls, one of many lovely cascades in Lamington NP. *Top inset:* Southern Forest Dragon.

A path leads through a stand of ancient Antarctic Beech.

More than 150 km of walking tracks lead through the park's rainforests, eucalypt forests, grassy woodlands and mountain heaths. The 22 km Border Track links Binna Burra to Green Mountains. There are exceptional views over the Numinbah Valley to Mt Warning from the summit region along the track. Most of the park's tracks branch off this main track to lookouts, waterfalls and rocky creeks.

The graded trails vary in length and difficulty, with some accessible to wheelchairs, so everyone can enjoy exploring at Lamington. There are several unmarked routes through the undeveloped southern region which are popular with fit and experienced bushwalkers.

Chalahn Falls in the green heart of Lamington National Park.

The treetop boardwalk at Green Mountains.

A crows nest fern grows on a looping vine.

Green Mountains offers a range of excellent rainforest walks and cliffside lookouts over the western valleys and escarpments. One of the best and shortest trails is a suspended boardwalk which offers an unusual look at the world of the treetops.

The Binna Burra section has its share of long and short rainforest walks. The Daves Creek circuit also takes in an interesting area of stunted eucalypt woodland and heath with banksias and bottlebrushes. A 5 km trek to the top of Coomera Gorge reveals a spectacular waterfall.

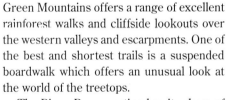

Coomera Falls drops uninterrupted for many metres.

LAMINGTON'S WILDLIFE

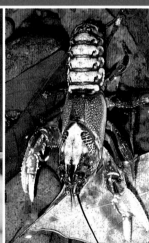

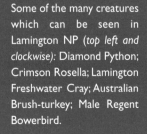
Some of the many creatures which can be seen in Lamington NP (*top left and clockwise*): Diamond Python; Crimson Rosella; Lamington Freshwater Cray; Australian Brush-turkey; Male Regent Bowerbird.

ACCESS

Boat or barge access from Rainbow Beach, Hervey Bay and River Heads. Suitable for 4WD only. Vehicle permit required. Landing strip for light aircraft at Orchid Beach. Tours operate from Noosa and Hervey Bay.

CAMPING

Coastal sites at Dundubara, Waddy Point and Wathumba. Forest and lake sites at Lake McKenzie, Lake Boomanjin, Lake Allom and Central Station. Beach camping permitted. Booking required and fees charged. Privately-owned accommodation available.

WHEN TO VISIT

Popular all year especially during school holidays. Bushwalking is best May–October.

SAFETY

Strong currents and rips make ocean swimming dangerous. Don't feed the Dingos — they are wild dogs and they bite!

Plants anchor the shifting dunes.

FRASER ISLAND

Fraser Island is a World Heritage Area of outstanding beauty, renowned for its idyllic lakes and majestic rainforests.

Fraser's eastern shores are a mecca for sightseers and fishing enthusiasts. High coastal dunes with stark sandblows and cliffs of multi-coloured sand rise above the long, wide beaches.

In the north of the island, the sandmass descends from its lofty heights through open forest, fragile heath and wetlands to Hervey Bay. In several places the watertable emerges between the low dunes forming broad shallow lakes.

Half of the world's perched dune lakes are found on Fraser. These unique lakes occur when rain collects above the watertable in depressions with hardened floors of decayed plants. The largest of these is Lake Boomanjin, its water stained by surrounding paperbark trees. Lake McKenzie, with its forest backdrop and white beach, is one of Fraser's prettiest lakes. The clear water reflects the changing colours of the sky.

Freshwater lakes, sand dunes and beaches create beauty on Fraser Island, one of Australia's World Heritage Areas.

Wanggoolba Creek runs through rainforest which is remarkable because it is growing in sand.

The dramatic cliffs of Indian Head shelter a sandy beach.

Pandanus and cliffs of coloured sand are island features.

A network of old logging tracks spreads over the island offering 4WD or walking access to coastal and inland features.

Camping and visitor facilities make Central Station an ideal base for inland day trips. The walk through to Pile Valley is one of the best rainforest experiences in Queensland.

The trail passes through a lush understorey of palms, ferns and lilly pillies alongside a crystal clear creek which flows silently over pure white sand. Beyond the creek banks, satinay, brushbox and kauri tower above heavily buttressed figs and quandong.

The island also has a rich Aboriginal heritage. The middens, artifacts and significant sites scattered over the island are reminders of the time when Fraser was a paradise known as K'gari.

ISLAND WILDLIFE

Fraser Island is home to many creatures. Most obvious are the Dingos *(above)*, which should not be fed or approached even if they appear friendly. The island's reptiles include the Lace Monitor *(left)*. The unique lakes *(below)* have their own fauna, including turtles and fish.

Vegetation clings to an isolated sand dune.

COOLOOLA

The eastern face of Cooloola stretches over 30 km between Teewah and Double Island Point before curving round to Rainbow Beach. Its steep frontal dunes feature coloured cliffs and sandblows.

The beach can be driven on by 4WD at low tide and, with easy access from either end, traffic gets fairly hectic. Cool lakes set amongst the rainforest at Freshwater offer a refreshing break from the beach.

Bushwalking and canoeing are great options for exploring inland. The 46 km Cooloola Wilderness Trail has four access points between Elanda and the Rainbow Beach Road. The 3 day walk takes in paperbark swamps, creekside rainforest, eucalypt forest and open heath.

Trails offering full- and half-day walks can be found at the camping and picnic sites along the upper Noosa River. The river is navigable by canoe for 29 km beyond the Kinaba visitor centre. Trees overhang the river as it winds through hidden channels and narrow reaches. The deep, tannin-stained water produces stunning early morning reflections.

While enjoying the beauty of this area, it is best to stick to the well-marked track, as off-track walking through dense heath and wetlands is really unpleasant.

As the river changes from brackish to fresh water, paperbarks remain a constant feature but mangrove trees and native hibiscus give way to banksia and floating sedge mats, then fringes of vine scrub. Eucalypt forest and heath plains dominate the landscape beyond Harry's Hut.

Paperbarks border this freshwater creek at Cooloola.

Fishing at sunrise on Cooloola Beach.

4WDs take anglers to the best surf-fishing spots on Cooloola Beach.

ACCESS

By boat from Noosa, Boreen Point and Elanda. Dry weather access to Harry's Hut for conventional vehicles via Cooloola Way. 4WD beach access from the Tewantin ferry and Rainbow Beach.

CAMPING

Vehicle-based camping at Freshwater and Harry's Hut. Limited numbers and length of stay at bush camp sites along the river. Permit required and fees charged. Book in advance.

WHEN TO VISIT

Any time of the year. Late August through September for the wildflowers.

SAFETY

Rips and strong currents make surf swimming dangerous. Rain brings out the mosquitoes and floods some riverside camp sites. Boil river and creek water before drinking.

Scaly-breasted Lorikeet on grevillea.

GIRRAWEEN NATIONAL PARK

Nature's stepping stones adorn a creek in Girraween NP.

Girraween lies within the fascinating landscape of Queensland's granite belt. There is a powerful yet whimsical nature to its smooth, domed summits and precariously perched boulders. In spring, these massive slabs of grey granite stand starkly against the riotous colours of surrounding heath and woodland flowers.

Girraween is a park the whole family can enjoy with its well-maintained facilities and walking tracks. It also offers several off-track challenges for experienced bushwalkers and climbers, including the walk through to Bald Rock NP.

Graded walking trails radiate from the visitor centre, providing easy access to impressive rock formations, delightful waterholes and boulder-strewn creeks.

Walks vary between a 500 m stroll to Granite Arch and the 5 km traverse of the park to the Mt Norman summit with its panoramic views. The final ascents to some of the summits are difficult, especially in wet conditions. However, there are plenty of opportunities for young or novice bushwalkers to try some rock scrambling.

The park's forests, woodlands, heaths and swamps are home to a variety of animals. Many are attracted to water and a walk to Dr Roberts Waterhole is always rewarding.

Granite formations invite exploration in Girraween.

Dawn and late afternoon are ideal times to look for grey kangaroos, wallaroos and wallabies feeding in open grassy areas. A spotlighting walk along the Granite Arch track often turns up nocturnal mammals such as possums, gliders and wombats. For a good cross-section of the park's birdlife, try the main track which crosses several habitats. Keep an eye open for the Superb Lyrebird in damp forest areas and the Turquoise Parrot which feeds on grasses and heath plants.

ACCESS

240 km SW of Brisbane via Stanthorpe and Ballandean on the New England Hwy.

CAMPING

Good facilities at Bald Rock Creek and Castle Rock. No powered sites. Booking and permit required. Fees charged. Overnight camping permit available.

WHEN TO VISIT

Any time of the year, but cold in winter. Wildflowers are best August–October. Good weather for bushwalking March through May.

SAFETY

Swimmers should beware of submerged objects and changing water levels in creeks. Granite rocks become very slippery when wet.

Eastern Grey Kangaroo.

ACCESS

Conventional vehicle access in dry weather. No through roads within the park. 4WD required for remote sections.

Carnarvon Gorge: 245 km N of Roma via Injune turn off to unsealed road at Wyseby or 105 km S of Rolleston.

Mt Moffat: 160 km NW of Injune via Womblebank or Westgrove; unsealed road.

Ka Ka Mundi: 120 km SW of Springsure.

Salvator Rosa: 172 km SW of Springsure or 197 km NE of Tambo.

CAMPING

Good facilities in creekside setting at Carnarvon Gorge. Bush camping at designated sites elsewhere. Booking and permit required. Fees charged.

WHEN TO VISIT

Late autumn—early spring. Winter nights cold; days ideal for walking.

SAFETY

Boil creek water before drinking.

Carnarvon Creek runs through Carnarvon Gorge, the most accessible part of Carnarvon NP.

Carnarvon NP covers 298 000 ha of the vast sandstone plateau of Queensland's central highlands.

Time and nature have carved the undulating tableland into an fascinating world of deep canyons, white-faced cliffs and sculpted outcrops, the eroded sandstone evoking images of ancient ruins and lost civilisations.

Carnarvon Gorge is the most popular and accessible section of the park with good visitor facilities and nearby lodge accommodation. The winding gorge penetrates 30 km into the plateau.

Its sheer cliffs rise over 100 m to the wooded foothills and escarpments of the Consuelo Tableland.

On the gorge floor, cabbage palms, she-oaks and weeping bottlebrushes edge the clear waters of Carnarvon Creek. Shrubs and tall eucalypts take the higher ground above the creek banks. The creek's numerous tributaries have carved narrow chasms in the gorge walls creating moist pockets where figs, tree ferns and other rainforest plants cluster. Orchids, ferns and mosses cling to the damp rock walls.

Carnarvon Creek has carved a gorge from sandstone.

Tree ferns grow from the walls of the gorge.

This green haven with permanent water attracts a diversity of birds, reptiles and mammals. A creekside walk at dawn is always rewarding when rock wallabies and grey kangaroos come to drink and the birds are most active. A night spotlighting walk in the forest usually reveals the eyeshine of possums, gliders and quolls.

The beauty and abundance of the gorge was not lost upon the Aboriginal people who occupied the region until the late 1800s. Caverns and rocky overhangs shelter stencilled, etched and freehand artwork depicting their daily concerns.

An extensive track system runs through the gorge providing a range of full- and half-day walks. The main track follows Carnarvon Creek 10 km upstream. Several short tracks branch off to side gorges, Aboriginal art sites and cliffside climbs to lookouts over the gorge.

The magic of Carnarvon extends beyond the gorge to the valleys, plains and escarpments on the plateau. Prominent bluffs and dramatic outcrops crown the plains and valleys. Many are Dreaming sites important to Aboriginal creation beliefs.

Creeks and springs flow from the land, feeding several of the State's river systems. A mosaic of woodland, eucalypt forest, brigalow and natural grassland cover the tableland. The diversity of plant habitats also includes small areas of softwood scrub.

Rough 4WD tracks provide separate access to the plateau at Ka Ka Mundi, Salvator Rosa and Mt Moffat. Self-sufficiency and a sense of adventure are needed to make the most of these undeveloped areas.

Carnarvon Gorge NP has many remarkable galleries of Aboriginal rock art.

Designated bush camping sites have no facilities but are located in shady settings beside rivers and waterholes. Tracks within each section lead to impressive sandstone features as well as excellent examples of Aboriginal rock art.

This is beautiful bushwalking country particularly during the cooler winter months. As surveyor Thomas Mitchell wrote of his 1846 visit: "It was a discovery worthy of the toils of pilgrimage".

Platypus are common in the park's waterways.

Picturesque sandstone formations such as this arch are features of Carnarvon Gorge.

GREAT BARRIER REEF MARINE PARK

ACCESS

Privately-owned, charter and tour boats from main departure centres:
Cairns/Port Douglas
Airlie Beach/Shute Harbour
Gladstone/Rosslyn Bay
Some islands accessible by air.

ACCOMMODATION

Ranges from self-sufficient bush camping on some national park islands to luxury resorts on privately-owned islands.

WHEN TO VISIT

Comfortable temperatures and relatively stable weather from April through October.

SAFETY

Be prepared for bad weather.
Avoid dangerous marine animals.

The Great Barrier Reef Marine Park stretches 2000 km along the Queensland coast between Fraser Island and the tip of Cape York. Within its tropical waters lies the world's largest system of coral reefs.

To the east, long ribbons and platforms of coral rise 200 m above the continental shelf, creating a barrier to the forces of the Pacific Ocean. Behind this great outer wall is an inner labyrinth of reefs, lagoons, sand cays and deep channels. Further inshore, coral reefs fringe the coastline and high continental islands.

The foundations of this vast empire are the millions of hard limestone skeletons secreted by generations of tiny animals known as coral polyps. A thin layer of these colourful animals forms the living, growing part of a reef.

While the bulk of a coral reef is not alive, there is an abundance of life in and around it. An underwater version of Alice's wonderland, the sheer numbers, colours and shapes of marine plants and animals defy the imagination.

About 1500 species of fish have been found on the Great Barrier Reef, ranging from delicate anemonefishes to the giant Manta Ray. The warm reef waters are also the winter breeding ground of Humpback Whales. In late spring the coral cays attract thousands of nesting seabirds and turtles, and migratory wading birds arrive from as far away as Siberia seeking food and rest.

Lady Musgrave Island, with its lagoon and encircling coral reef.

With over 2900 coral reefs and some 900 islands and cays, this World Heritage Area also attracts thousands of tourists. A variety of Great Barrier Reef experiences is on offer, focusing mainly on the inner reefs and continental islands. Some visitors come in their own craft, while others stay at island resorts. Camping is allowed on some islands (apply for permit well in advance).

THE SOUTHERN SECTION

The Mackay/Capricorn section of the park stretches from north of Bundaberg to the Whitsundays and extends seaward some 300 km at its widest point. It includes Lady Elliot Island, Heron Island, the Bunker Group and the Capricorn Group. A number of cays in this section of the Reef are national parks. Reef walking, turtle watching and superb diving are on offer as well as the usual water sports.

There is something to see all year here, including whale-watching in June–August and turtle-hatching in January–April.

Guided tours of the reef flat are features of most resorts.

Scenes from Whitsunday NP: *Top left and clockwise:* Whitehaven Beach; Nara Inlet; Hill Inlet; a safe harbour; Pentecost island.

LIFE ON THE REEF

Top left and clockwise: Long-armed Sea Star, one of many sea stars on the reef; a Pink Anemonefish sheltering in anemone tentacles; a Green Turtle swimming over the reef; bannerfish swimming along a coral outcrop.

THE CENTRAL SECTION

The Cumberland Islands and the more northerly Whitsundays present a completely different aspect of the marine park. There are around 70 islands in the Whitsunday Group, all but a few of them national parks. They are not coral cays, but the peaks of drowned mountains, and their fringing reefs are rich in marine life. Many are forested and have plentiful wildlife, and there are holiday resorts on a few. They can be explored on tours, or in a hired cruise craft, known as "bareboating". Bareboat charter companies can be found at Airlie Beach, mainland centre for the Whitsundays.

Sunset in the Whitsundays.

About 70% of Magnetic Island, near Townsville, is national park. It has beaches, bushwalking tracks and plenty of birds. While at Townsville, it is worth visiting the Great Barrier Reef Wonderland, for easy viewing of underwater life.

THE NORTHERN SECTION

Hinchinbrook Island is Australia's largest island national park, 399 km^2 in area and with Mt Bowen rising to 1121 m. It is separated from the mainland by the narrow Hinchinbrook Channel. Reached from the mainland town of Cardwell, the island has dense tropical rainforest, beaches and bays and extensive mangrove forests. The whole island is national park, and has camping areas as well as the Cape Richards resort on the north of the island.

There are short walks around the resort, a longer walk from Macushla Bay to Shepherd Bay, and a wilderness walk over 30 km from Ramsay Bay to Zoe Bay.

This walk takes two strenuous days or three pleasant ones over beaches, mountains, rainforests and creek crossings. Fitness, water supplies, sunscreen and insect repellent are essentials for achieving all (or part) of this rewarding walk.

A graceful craft ready for Barrier Reef adventure.

Mangrove seeds will drop into the sea for dispersal.

Mangroves and tidal creeks on Hinchinbrook Island.

Hillock Point is in the foreground of this view over Hinchinbrook Island's spectacular terrain.

It is possible to dive, snorkel or just enjoy the reef from boats which make daytrips from Cairns and Port Douglas.

Exploring Agincourt Reef from a dive boat.

Off the coast from Tully lie Dunk, Bedarra and the Family Islands. Three-quarters of Dunk and most of the other islands are national parks. All of them are covered in rainforest and they all experience high rainfall. People who visit during the December–March wet season should be prepared for this weather.

The Low Isles, Green Island and low-lying Michaelmas Cay are popular one-day excursions in the Cairns region and provide a good introduction to the marine and terrestrial life of sand cays. Fitzroy Island is a continental island, a popular daytrip destination accessed from Cairns. Lizard is the furthest north of the Barrier Reef islands (240 km north of Cairns) to offer resort facilities. Lizard and neighbouring islands are national parks, and since they are only 15 km from the outer edge of the reef they offer superb diving and snorkelling. Lizard was named after the monitor lizards, or goannas, seen by Captain Cook in 1770 and still present today.

Raine Island, at the northern extremity of the Reef.

Green Island is 27 km north-east of Cairns.

The Low Isles consist of Low Island, with its 1878 lighthouse, and mangrove-covered Woody Island.

WOOROONOORAN NATIONAL PARK

The Dainty Green Tree-frog lives in the tree canopy.

Wooroonooran is the gem of the wet tropic national parks, protecting a large continuous area of rainforest. Bellenden Ker Range is its most spectacular facet being the highest, wettest coastal range in north Queensland.

The tributaries of the Mulgrave, Russell and Johnstone Rivers cut deeply into the range creating cascades and waterfalls as they descend from its double summit.

The park's basaltic soils support several types of tropical rainforest, from the highland thickets of windswept peaks to the complex lowland forest of deep river valleys.

ACCESS

Bellenden Ker section: 8 km sealed road from the Bruce Hwy 28 km N of Innisfail. Gourka Rd via Lamins Hill Lookout 15 km E of Malanda.

Palmerston section: 35 km W of Innisfail on the Palmerston Hwy.

CAMPING

Basic facilities at Henrietta Creek. No camping at Josephine Falls. Overnight bush camping permits available for the Summit and Goldfields trails.

WHEN TO VISIT

Spectacular during the wet season but more comfortable April–October

SAFETY

Be careful of slippery rocks around waterfalls and rockpools. Bushwalkers should register with the park ranger before setting out on the summit trail.

Nandroya Falls, one of Wooroonooran NP's many glorious cascades.

Ulysses Butterfly. (SB)

The Herbert River Ringtail Possum lives in rainforest.

The cooler, wetter rainforest of the ranges are a refuge for Herbert River and Green Ringtail Possums and Lumholtz's Tree-kangaroo. These nocturnal marsupials feed on the leaves of trees.

Some of the larger birds found between the forest floor and high canopy include the Satin and Golden Bowerbirds, Victoria's Riflebird and the Tooth-billed Catbird.

The park's many creeks are home to Platypus, rainbow fish and freshwater turtles. There are also many spectacular butterflies, including the magnificent green and gold Cairns Birdwing and the iridescent blue Ulysses Butterfly.

Three routes traverse Wooroonooran from east to west offering a rare opportunity to experience the progression of rainforest from lowland to highland by car or on foot over some walks of varying difficulty.

The Palmerston Highway follows a rainforested spur of the Bellenden Ker Range as it climbs from the coastal plain to the Atherton Tableland. Tributaries of the North and South Johnstone Rivers spill into deep river valleys either side of the highway.

A series of walking tracks ranging from 800 m to over 7 km descend the north face of the spur between Crawford's Lookout and the Henrietta Creek camping area. The trails follow parts of Douglas and Henrietta Creeks through luxuriant rainforest providing access to lookouts, waterfalls and rockpools.

The Josephine Falls Trail, an easy 800 m walk through lush lowland rainforest, leads to the rockpools at the base of Josephine Falls. The mountain-fed waters of Josephine Creek surge through a series of waterfalls and smooth granite rockpools before joining the Russell River on its journey to the coastal plains.

The rugged and challenging Bartle Frere Summit Trail climbs the Bartle Frere rock mass to a series of cloud shrouded peaks. Experienced bushwalkers can approach the ascent as a 15 km return trip from either Josephine Falls or Gourka Road or as a 15 km traverse between the two.

The Johnstone River runs from the summit of the Great Dividing Range through rainforest towards the Coral Sea.

This ungraded but well-marked trail passes through dense lowland rainforest, open scrub and stunted highland forests as it follows creek gullies and rocky ridges. There are outstanding views of the Atherton Tableland, coastal plains and the Bellenden Ker summit along the way.

The Goldfields Trail is a 19 km one-way walk which retraces an historic gold prospecting trail to the north-western slopes of Bartle Frere. The trail starts and finishes about 1 km outside the park's east and west boundaries. It can be accessed from the Boulders Scenic Reserve which is 7 km west of Babinda, or from the Goldsborough Valley State Forest which is 15 km east of the Gillies Highway.

This is a long but comfortable walk through lowland rainforest over the low saddle between the Bellenden Ker and Bartle Frere summits. The rainforest changes significantly along the track as a result of logging, cyclones and increasing altitude.

Tchupala Falls can be reached along an easy walking trail.

Rainforest is a magical place to explore.

ACCESS

Sealed road access 80 km N of Cairns via Mossman.

CAMPING

No camping at Mossman Gorge. Permit available for overnight bush camping in remote areas.

WHEN TO VISIT

Accessible any time of the year. April to October offer comfortable walking conditions.

SAFETY

Be aware of submerged objects and strong currents in the river. Bushwalkers going beyond the gorge require a permit.

Daintree covers more than 75 900 ha of the Wet Tropics World Heritage Area. It is a showcase of nature's talents ranging from the dramatic to the sublime.

To the west are the cloud-capped peaks and highland-rainforest-covered slopes of the Great Dividing Range. The Mossman and Daintree Rivers make dramatic descents from this wild country, cascading over boulders and through deep valleys to the coastal plain.

To the east is a sublime coastline where lush lowland rainforest grows within metres of fringing coral reefs. There are few other places on Earth where these two complex habitats meet so closely.

Plant and animal life in the park's rainforests include the rare and the unexpected. Cassowaries 2 m tall roam the coastal forest floor while possum-like tree-kangaroos climb among the tree branches. Primitive species such as the Musky Rat-kangaroo and the *Idiospermum* plant offer clues to the evolution of life in Australian rainforests over millions of years.

The highlands are so remote and rugged that much of the park is inaccessible except to the well-equipped and experienced bushwalker. There is, however, car access to short walking tracks and boardwalks where visitors can experience lowland rainforest, palm swamps, mangroves and superb coastal scenery.

The Mossman River flows swiftly from lush rainforests across the coastal plain into the Coral Sea.

A crystal stream in Daintree NP.

Boulders along a rainforest stream are covered with mosses.

Moss covers boulders in a mountain stream in the high country of Daintree NP.

Unique Fan Palm forests are a feature of Daintree NP.

LIFE IN THE RAINFOREST

Above and clockwise: Cooktown Orchid; Green Ringtail Possum; Male Southern Cassowary; Cairns Birdwing Butterfly.

MOSSMAN GORGE

This remote western section is a wilderness of tropical rainforest, rugged scarps, deep river valleys and mountain creeks.

The circuit track at Mossman Gorge provides a good introduction to lowland rainforest in a setting of rockpools and cascades. Signs along the self-guiding trail identify many of the trees and explain their importance to native animals and humans.

Where the Mossman River and its tributaries have created openings in the forest canopy, the understorey is dense with palms, shrubs and young trees taking advantage of the sunlight. Behind these, thick vines, ferns and other such plants, called epiphytes, cling to the tree trunks.

Daintree rainforest.

23

ACCESS

Tours from Cairns.

The Daintree ferry is 104 km N of Cairns via Mossman. There is a partly sealed, narrow road from the ferry to the Cape. The 4WD Bloomfield track is impassable after heavy rain. Unsuitable for caravans.

Snapper Island, at the mouth of the Daintree River, is accessible by privately-owned and tour boats.

CAMPING

Basic facilities at Noah Beach. Permit required and fees charged. Bush camping at Snapper Island.

WHEN TO VISIT

April to October, in the Dry, is the best time.

SAFETY

Crocodiles present in creeks, rivers and estuaries at all times. Be aware of marine stingers October–April.

CAPE TRIBULATION

Sorrow and tribulation were Captain Cook's lasting impressions of the Daintree after his ship encountered one of its fringing reefs. In more favourable circumstances he would have been the first of many visitors to marvel at Cape Tribulation's remarkable beauty and diverse lifeforms.

The Alexandra lookout offers sweeping views of the coast. To the south, the Daintree River meanders its way through mangrove forests towards the sea and Snapper Island. Stretching northwards, the forest-clad slopes of Thornton Peak rise above coral reefs to a series of ragged peaks. The lush coastal forest becomes a dense, wind shorn carpet on the mountain peaks and rocky ridges. No formal tracks ascend the ranges and off-track navigation is extremely difficult.

Daintree NP's narrow coastal section protects the country's largest area of tropical lowland rainforest which features unusual palm swamps.

The area supports many bird species, including Southern Cassowaries, Noisy Pittas and migratory kingfishers.

The 800 m Marrdja walk at Oliver Creek provides a good introduction to this habitat as well as to the mangroves fringing Noah Creek. Three short walks at the cape, taking in rainforest, mangroves and coastal views, can be incorporated into a 4-hour return walk between north Myall Beach and Emmagen Creek.

The largest group of fringing coral reefs in eastern Australia follows the shoreline between Snapper Island and Bloomfield. The waters are relatively clear of sediments and there are opportunities for kayaking and snorkelling with easy access from the beach.

Rainforest trees may bear flowers direct on their trunks.

Looking through rainforest to Barrier Reef waters.

Cape Tribulation, where the rainforest meets the reef.

LAWN HILL NATIONAL PARK

Lawn Hill, set in the remote north-western highlands of Queensland, is an area of surprising contrasts. Beneath the vast plateau of sunbaked limestone, freshwater springs feed the park's gorge-carving creeks. Figs and palms crowd around deep waterholes below the gorge cut through the sandstone ridges.

There are good facilities and a 20 km walking track system at Lawn Hill Gorge. Graded trails lead to scenic lookouts, Aboriginal cultural sites and unusual geological formations. Off-track walking on the plateau is not difficult, but can be uncomfortably hot during the middle of the day.

It is an enjoyable canoe trip or a 4 km walk to Island Stack at the end of the gorge. Lawn Hill and Widdalion Creeks have eroded this 50 m sandstone column. Other features along the creeks include a natural spa, Indarri Waterfall and tufa deposits of spongy limestone.

Etched and painted rock shelters at the gorge speak of a long and fascinating association between Aborigines and the country. A shady 4.5 km track provides access to Wild Dog Dreaming where signs explain the site's significance. An even older history lies embedded in limestone 65 km south at Riversleigh, where fossil deposits record 20 million years of change in landscape and fauna.

Lawn Hill Creek is fed by freshwater springs.

Dawn, dusk and night spotlighting walks are the best times to look for the abundant bird and animal life of the park. Wallaroos, tiny bats and ringtail possums hide among rocky outcrops and overhangs during the day. The Olive Python, which can grow to an impressive 6.5 m in length, is another rock-dwelling night-feeder.

Sunrise brings Sandstone Shrike-thrushes whose melodious calls echo through the gorge. Daylight also reveals some of Lawn Hill's living fossils. The tranquil gorge waters are home to Freshwater Crocodiles and Northern Snapping Turtles.

Lawn Hill Gorge is an oasis in the arid plains of north-western Queensland.

ACCESS

220 km SW of Burketown via Gregory Downs on an unsealed road. Dry weather access. 4WD recommended.

285 km NE of Camooweal via Riversleigh for 4WD only.

Light aircraft landing strip at Adel's Grove.

CAMPING

Well maintained facilities for self-sufficient campers. Bookings essential, permit required and fees charged.

WHEN TO VISIT

Avoid summer heat by visiting May to October. Winter nights are cold but days are ideal for walking.

SAFETY

Carry water when travelling beyond the gorge area. Take care when canoeing or swimming. Freshwater crocodiles are harmless unless provoked.

Common Wallaroo.

ACCESS

80 km track through private property from Birdsville to eastern boundary. 4WD only and permit to traverse is required. Signposted track follows QAA survey line for 130 km to Poeppel Corner. Desert Parks Pass required if travelling through to SA.

CAMPING

Self-sufficient bush camping only. Permit required and fees charged. Take all supplies.

WHEN TO VISIT

July to September to avoid extreme heat of summer. Temperatures are freezing at night but rain less likely.

SAFETY

Make sure vehicle is in good condition before entering park. Travel with at least one other vehicle and inform Birdsville police of itinerary. Carry plenty of water. Take equipment for digging vehicle out of sand if necessary.

Central Netted Dragon.

Queensland's largest national park of over 1 million hectares occupies just a small corner of the Simpson Desert. This vast arid land possesses a simple grandeur.

Stately ranks of rust coloured dunes stretch towards the heart of Australia. It takes skill and patience to negotiate the difficult terrain of the desert country. The gentle western dune slopes give no warning of the unstable crests of live sand or the 10 to 30 m drop from the steep eastern faces. Drifting sand, claypans, gibber plains of hard rounded ironstone and saltpans of sticky mud lie in the shallow valleys between the rows of dunes.

The desert is isolated and uncompromising. Its temperatures are extreme and its rainfall meagre. Yet a surprising variety of plants and animals have adapted to these harsh conditions. Scattered mulga and gidgee trees take the low ground along with grasses and shrubs such as grevillea, hakea, emu bush and cassia. Large spiky clumps of spinifex dominate the western slopes, while canegrass and succulents cope with the shifting crests and dune faces.

Rain, when it does fall, transforms the desert. Masses of wildflowers sprout from dormant seeds and claypans become temporary waterholes attracting flocks of birds from other regions.

Desert animals can be difficult to find. Many are nocturnal, others well camouflaged, but most leave tracks in the sand.

Clumps of spinifex provide refuge for desert creatures.

Small mammals forage at night on plants, seeds or insects which give them enough water to survive. They take refuge in burrows, beneath spinifex clumps or in the deeply cracked earth during the day. Snakes and lizards move between sun and shade and avoid the hottest part of the day.

Simpson Desert is definitely a park for the adventurous and well prepared traveller. There are no roads, walking tracks or facilities. There is, however, solitude and unique beauty in this remote wilderness.

The few hardy shrubs which grow on the sand dunes flower after rain has fallen on the desert.

NEW SOUTH WALES

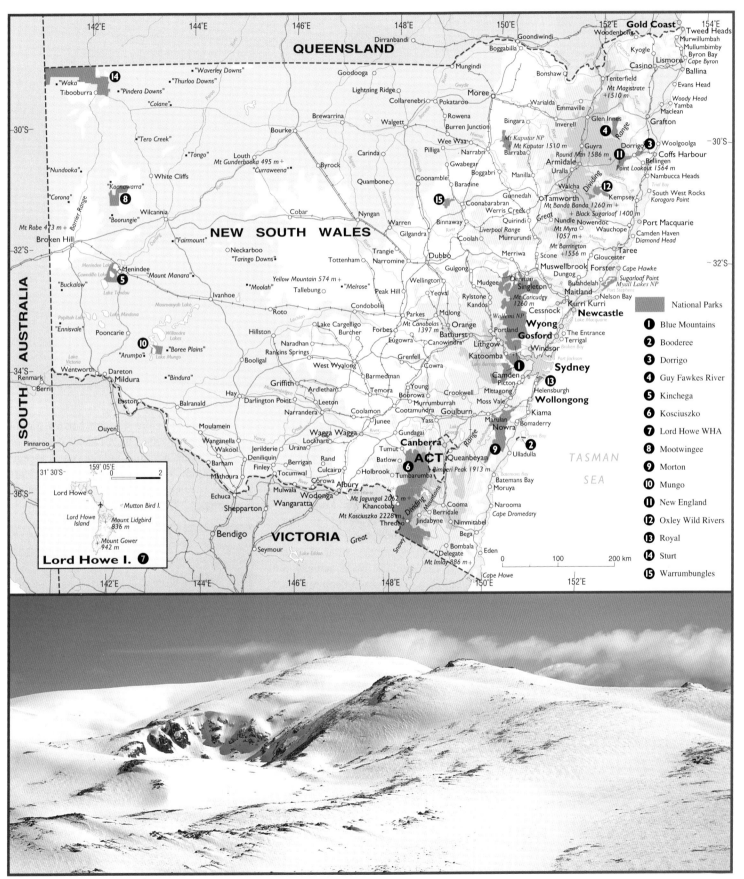

QUEENSLAND

Dirranbandi
Goondiwindi
Gold Coast
Woodenbong
Tweed Heads
Boggabilla
Kyogle
Murwillumbah
Mullumbimby
Mungindi
Bonshaw
Casino
Byron Bay Cape Byron
Lismore
Goodooga
Tenterfield
Ballina
"Waverley Downs"
Lightning Ridge
Moree
Warialda
Mt Magistrate +1510 m
Evans Head
"Waka"
"Thurloo Downs"
Collarenebri
Pokataroo
Emmaville
Woody Head
Tibooburra
"Pindera Downs"
Rowena
Bingara
Glen Innes
Yamba
"Colane"
Brewarrina
Walgett
Burren Junction
Inverell
Range
Maclean
Wee Waa
Narrabri
Guyra
Dorrigo
Woolgoolga
Bourke
Pilliga
Barraba
Round Mtn 1586 m
Coffs Harbour
"Tero Creek"
"Tongo"
Louth
Byrock
Carinda
Gwabegar
Manilla
Walcha
Point Lookout 1564 m
Bellingen
Mt Gunderbooka 495 m +
Quambone
Boggabri
Uralla
Nambucca Heads
"Nundooka"
"Curraweena"
Baradine
Gunnedah
Tamworth
Mt Banda Banda 1260 m +
Kempsey
South West Rocks
"Corona"
White Cliffs
Coonamble
Werris Creek
+ Black Sugarloaf 1400 m
Korogoro Point
"Koonawarra"
Wilcannia
Cobar
Nyngan
Coonabarabran
Quirindi
Nundle Nowendoc
Port Macquarie
"Boorungie"
Warren
Gilgandra
Liverpool Range
Murrurundi
Wauchope
Camden Haven
Mt Robe 473 m +
NEW SOUTH WALES
Coolah
Mt Myra 1057 m +
Diamond Head
Broken Hill
"Fairmount"
Trangie
Narromine
Dubbo
Merriwa
Scone +1556 m
Gloucester
Taree
"Taringo Downs"
Tottenham
Mudgee
Muswellbrook
Forster Cape Hawke
Menindee
"Mount Manara"
Gulgong
Dungog
Myall Lakes NP
"Buckalow"
Yellow Mountain 574 m +
Denman
Bulahdelah
Singleton
Nelson Bay
"Moolah"
Tallebung
Peak Hill
Wellington
Rylstone
Kandos
Mt Coricudgy 1260 m
Maitland
Kurri Kurri
"Ennisvale"
Ivanhoe
Yeoval
Cessnock
Newcastle
Roto
Condobolin
Parkes
Molong
Orange
Wollemi NP
Wyong
National Parks
Pooncarie
Lake Cargelligo
Forbes
Mt Canobolas 1397 m
Portland
Gosford
1 Blue Mountains
Hillston
Burcher
Eugowra
Canowindra
Lithgow
Windsor
The Entrance
2 Booderee
"Arumpo"
Naradhan
Grenfell
Bathurst
Katoomba
Terrigal
3 Dorrigo
"Boree Plains"
Rankins Springs
Cowra
1
Sydney
4 Guy Fawkes River
Wentworth
Booligal
West Wyalong
Young
Camden
Picton
13
5 Kincega
Dareton
"Bindura"
Griffith
Ardlethan
Temora
Boorowa
Crookwell
Helensburgh
6 Kosciuszko
Mildura
Hay
Darlington Point
Leeton
Murrumburrah
Moss Vale
Wollongong
7 Lord Howe WHA
Balranald
Narrandera
Cootamundra
Goulburn
Kiama
8 Mootwingee
Moulamein
Coolamon
Junee
Yass
Bomaderry
9 Morton
Wanganella
Wakool
Wagga Wagga
Gundagai
Marulan
Nowra
10 Mungo
Barham
Jerilderie
Urana
Lockhart
Lake
Bomaderry
11 New England
Deniliquin
Finley
Rand
Tumut
Batlow
Canberra
Queanbeyan
Ulladulla
12 Oxley Wild Rivers
Mathoura
Berrigan
Culcairn
Holbrook
ACT
Bimberi Peak 1913 m
13 Royal
Echuca
Tocumwal
Corowa
Albury
Tumbarumba
Batemans Bay
14 Sturt
Mulwala
Wodonga
Mt Jagungal 2062 m +
Cooma
Moruya
15 Warrumbungles
Shepparton
Wangaratta
Khancoban
Berridale
Narooma
Bendigo
Mt Kosciuszko 2228 m +
Jindabyne
Nimmitabel
Cape Dromedary
Thredbo
Bega
Seymour
Bombala
Eden
Delegate
Mt Imlay 886 m +
Cape Howe

SOUTH AUSTRALIA
Renmark
Berri
Pinnaroo
Ouyen

VICTORIA

TASMAN SEA

Lord Howe
Mutton Bird I.
Lord Howe Island
Mount Lidgbird 836 m
Mount Gower 942 m
Lord Howe I. 7

Kosciuszko National Park includes Australia's highest mountain.

BLUE MOUNTAINS NATIONAL PARK

The Three Sisters overlook the Jamison Valley and spectacular areas of the Blue Mountains NP.

ACCESS

Car, train and bus. Most easterly area of park starts 60 km W of Sydney. Numerous access points from railway, Great Western Hwy and Bell's Line of Road. Fee charged for park use at Glenbrook. Southern section accessible from the old Oberon–Colong Stock Route.

CAMPING

Facilities at Murphy's Glen, Ingar and at Perrys Lookdown. Bookings required and fees charged at Euroka Clearing near Glenbrook. Restrictions apply to bush camping.

WHEN TO VISIT

Any season.

SAFETY

Creek water is unsafe to drink even when boiled.

Walkway to the Three Sisters.

The Blue Mountains presented Sydney settlers with a deceptive eastern face, veiling its true nature in a haze of blue. Its treed slopes appeared to sweep gently upwards from the coastal plain to a flat tableland. It appeared that this long, low range would be no impediment to the colony's westward expansion.

This illusion was dispelled when it was discovered that the foothills were actually a thickly forested maze of spurs and deep ravines leading to sheer-walled chasms and cliffs of bare-faced sandstone. It took some 25 years of trial and error before the colonists established the route over such a formidable barrier.

This section of the Great Dividing Range, where the crossing is, and its river valleys – the Jamison, Grose and part of the Megalong – now form the Blue Mountains NP.

Visitor facilities and walking tracks are concentrated along the edge of the plateau between Glenbrook and Mt Victoria, with the valleys' remaining wilderness areas accessible only on foot.

Clifftop lookouts and easy walks, including a wheelchair-friendly one, offer sweeping views of the remarkable features which innumerable creeks and rivers have carved out of the plateau's soft golden sandstone and shale. Other tracks

zigzag down the cliffs through narrow gorges and under waterfalls. The distances covered on these walks are not great but the return ascents can be tiring. Many of these valley tracks are linked, providing an opportunity for overnight bushwalks.

The mountain devil, common in the Blue Mountains bush.

The Superb Lyrebird is at home in the park's deep gullies.

The Park's impressive landscape can distract from its wealth of plant and animal life. Heath and open woodland are characteristic of the less fertile soils on the plateau and higher ridges. Tableland Road to McMahon's Lookout passes through both these habitats. It's a delightful drive in spring or early summer when boronia, scarlet waratah, ground orchids, and pea bush are in flower. Watch for honeyeaters and colourful rosellas along the way to the lookout over Lake Burragorang and the remote southern section of the park.

Below the edge of the plateau, cool damp forests of tall gum trees follow creeks through the lower slopes and valleys. Narrow gullies protect pockets of warm temperate rainforest filled with ferns, mosses, and vines. These wet forests are the haunt of the Superb Lyrebird, Golden Whistler, Satin Bowerbird and Rufous Fantail. Platypus, water-dragons and skinks are often seen along the creek banks.

A quiet moment in Blue Mountains wilderness.

This lookout offers views of Kings Tableland.

Govetts Leap, site of a legendary bushranger's death-leap.

Blue Gum forest in the Grose valley.

FALLS AND CASCADES

There are many waterfalls and cascades in the Blue Mountains, providing visions of silvery beauty as they tumble over sandstone ledges into pools surrounded by ferns and mosses. Wentworth Falls *(above left)* Beauchamp Falls *(above)* and Leura Cascades *(left)* are three of the mountains' loveliest landmarks.

The Neates Glen trail through the Grand Canyon to Evan's Lookout is a popular and exciting forest walk with rocky overhangs, fern-filled gullies and narrow chasms. The 5 km return walk from Perry's Lookdown to the valley floor is another good forest walk. Near the junction of Govetts Creek and the Grose River, dense eucalypt forest gives way to an unusual stand of pure blue gums.

For an easier walk through rainforest typical of the Blue Mountains, try the 3 km creekside trail from Euroka Clearing to the Nepean River. It is also worth the drive to Mt Wilson where a more complex type of rainforest grows on rich volcanic soils.

Discovering the park's wild side involves going beyond the graded track system. Those with abseiling and caving experience can take up the challenge of Claustral Canyon below Lightning Ridge. Experienced bushwalkers seeking wild, unspoilt country can descend Narrow Neck Plateau to the junction of the Coxs and Kowmung Rivers.

The walk to Yerranderie Historic Town via Kanangra Walls can take up to four or five days. This unmarked but well-known route can also be approached from the old Oberon–Colong Stock Route.

OXLEY WILD RIVERS NATIONAL PARK

ACCESS

By car. 5 visitor areas lie between 20 and 100 km E of Armidale via Waterfall Way. Access Southern gorges from Oxley Hwy E of Walcha.

CAMPING

No fees for camping at Wollomombi, Dangars and Apsley Gorges, Tia Falls, Long Pocket or Budds Mare. Booking fees for accommodation at East Kunderang Homestead. Booking required for camping at Riverside. No camping at Gara Gorge. Conditions apply to overnight bush camping.

WHEN TO VISIT

Bushwalking is most comfortable during spring and autumn. Waterfalls are at their best during spring and summer rains (September–February).

SAFETY

Take care along gorge edges and watch for submerged objects when swimming or canoeing.

Apsley Falls, a highlight of the park.

A wild river flows free down one of the Oxley Wild Rivers NP's picturesque gorges.

Over 500 km of creeks and rivers flow through this large segmented park located within easy reach of Armidale.

Free from the demands of electricity and water supply, the tributaries of the Apsley and Macleay Rivers plunge from the tableland into dramatic gorges. Flanked by narrow spurs, the rivers zigzag through the gorges to rendezvous beneath Carrai Plateau before emerging into a broad river valley.

Casuarinas and bottlebrushes fringe the wider river reaches. Open woodlands of eucalypt, cypress pine, wattle, and grasstree climb the rocky spurs and spread over the plateau.

A majestic eucalypt clings to a mountainside.

Turbulent mountain water cascades down Yarrowitch Falls into a pool surrounded by lichen-covered boulders.

Sheltered gullies and valley floors harbour large tracts of dry rainforest within this World Heritage Area. Dry rainforest is an unusual kind of rainforest where vines, ferns and epiphytes grow below a semi-evergreen canopy of lacebarks, kurrajongs, figs and giant stinging trees, often exploiting the ground-water in the gullies rather than relying on rain.

Visitor facilities and walking tracks focus on the waterfalls and rocky gorge heads north and south of the park's large wilderness area. At Dangars Gorge in the north, walking trails rim the gorge and descend to the Macleay River. Dangars, along with Gara and Wollomombi Gorges are favoured rock climbing sites. Wollomombi also offers a 2-day gorge walk to Long Pocket which takes in the Chandler River and rainforest.

Riverside and East Kunderang are ideal bases for canoeing expeditions, but access is restricted to 4WD vehicles.

A gorge carved by one of the park's wild rivers.

WILDLIFE OF THE GREAT DIVIDE

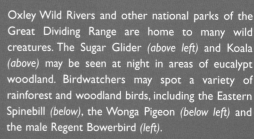

Oxley Wild Rivers and other national parks of the Great Dividing Range are home to many wild creatures. The Sugar Glider *(above left)* and Koala *(above)* may be seen at night in areas of eucalypt woodland. Birdwatchers may spot a variety of rainforest and woodland birds, including the Eastern Spinebill *(below)*, the Wonga Pigeon *(below left)* and the male Regent Bowerbird *(left)*.

ACCESS

Main entrance is from Dome Rd off the Dorrigo–Bellingen Rd 2 km from Dorrigo or 25 km from Bellingen. 10 km scenic drive along Dome Rd to Never Never section. Killungoondie Plains accessible from Slingsbys Rd heading SE from Megan.

CAMPING

No camping facilities in the park. Conditions apply to bush camping in Killungoondie Plains section.

WHEN TO VISIT

Any season. Autumn and spring are most comfortable for walking.

SAFETY

There are no barriers along the clifftop trails or around slippery rocks beside waterfalls. Leeches are common.

A view from under the waterfall shown opposite.

Dorrigo NP is a popular World Heritage Area offering an unforgettable introduction to the rainforests of central-eastern New South Wales.

This small park straddles the basalt edges of the Dorrigo Plateau where high rainfall gives rise to the Bellinger River's northern tributaries.

Rain and other agents of erosion have stripped layer upon layer of hardened lava from the plateau, feeding rich volcanic material into its many creeks. These fertile soils have accumulated in narrow gullies and deep valleys, fostering the growth of subtropical rainforest.

In the warm, humid conditions of these sheltered places, bangalow and walking stick palms compete for space and sunlight with tangled vines and epiphytic ferns and orchids. Strangler figs, giant stinging trees, flame trees and buttressed booyong and carabeen form a closed canopy high above the forest floor. These rainforest plants merge with eucalypt species as the terrain rises from low-lying gullies to drier, more exposed ridges clothed with forests of tall gum trees.

Visitors can get a bird's eye view of the volcanic rimmed plateau and its forested spurs and valleys from the Skywalk at the visitor centre near the park's main entrance. This raised boardwalk passes over the forest canopy to a viewing platform on the edge of the escarpment.

A 1 km road links the Dorrigo Rainforest Centre to the Glade picnic area where a second boardwalk affords a look at the birds of the rainforest canopy. As a follow-up, take a short stroll around the Satinbird circuit to experience the dimly lit forest floor.

Walking tracks provide access to the rainforest.

The Australian King-Parrot is common in the park.

While some visitors dally by a calm pool, others set off along a walkway to explore the rainforest.

Waterfalls are a feature of Dorrigo NP.

The 5.8 km Wonga circuit leads through rainforest to the foot of Crystal Shower Falls and a basalt overhang curtained by vines and falling water. It continues beneath the escarpment to a ridge of wet eucalypt forest with coastal views from Hardwood Lookout. The trail traverses the head of another rainforested valley past Tristania Falls and its hexagonal basalt columns before looping back to the Glade.

It's a 10 km scenic drive from the visitor centre to the warm temperate rainforests of the Never Never area. This is a simpler type of rainforest with fewer kinds of plants growing in a two-tiered structure dominated by coachwood, sassafras and crabapple. Rare and beautiful Dorrigo waratah grow in these forests. The plateau's poorer clay soils also support dense stands of eucalypt forest.

There are several good day walks with clifftop views which follow creeks to waterfalls above the escarpment. One side track zigzags down to a patch of unusual dry rainforest (relying more on ground-water than rain) at the base of Cedar Falls.

The differences in altitude and vegetation mean Dorrigo is home to a wide variety of wildlife. Ground-dwelling birds living in rainforest areas include lyrebirds, brush-turkeys, and Noisy Pittas, and the canopy is filled with Regent and Satin Bowerbirds, Wonga Pigeons, Green Catbirds and Australian King-Parrots.

The mammals are mostly nocturnal gliders and possums, although pademelons may be seen during the day. There are also some interesting reptiles, including the Land Mullet, one of the largest skinks.

A moss-covered forest giant lying on the forest floor will be home to many small rainforest creatures.

A crows nest fern grows on a strangler fig.

ACCESS

Conventional vehicle access on 15 km gravel road from turnoff on Armidale–Ebor–Grafton Rd 70 km E of Armidale. Road within park unsuitable for caravans.

CAMPING

No fees for basic camping facilities at Thungutti. Conditions apply to overnight bush camping. Bookings required for accommodation at the Residence, the Chalet and Tom's Cabin.

WHEN TO VISIT

Any time of year. Winter nights are cold; summer is mild and wet. Wildflowers are best from late August through September.

SAFETY

Take warm clothing and wet weather gear.

A picnic area in the park.

New England NP descends from the misty heights of its volcanic rimmed plateau to a cloistered world of richly forested ranges and valleys. It is one of several World Heritage rainforest parks scattered along the eastern edge of the Great Dividing Range between Queensland's Scenic Rim and Werrikimbe NP. New England's international importance extends beyond the rainforests to its outstanding scenic beauty and diversity of plant and animal life.

Point Lookout, 3 km inside the park, is the best place to appreciate the physical grandeur of this ancient landscape. Views from the prominent basalt headland sweep along the escarpment to the region's volcanic centre and out over the Bellinger River valley to the South Pacific Ocean.

Heath and snow gum woodland edge the cold, windswept cliffs. On the plateau behind, grassy eucalypt forests, swamps and heath provide homes for gliders, kangaroos, wallaroos, and a variety of reptiles. New England NP is also home to the rare Sphagnum Frog.

Lichens decorate the trunks of high-altitude trees in the park.

Cool temperate rainforests of Antarctic beech just below the escarpment give way to ridges of wet eucalypt forest and sheltered valleys of subtropical rainforest. Some of the inhabitants of the forest in the Bellinger River valley include bent-wing bats, Parma Wallaby, Leaf-tailed Gecko and Long-nosed Potoroo.

Storm clouds hang over the magnificent mountain scenery of New England NP.

Twenty kilometres of linked walking tracks lead through contrasting habitats above and below the escarpment. The 3 km Eagle's Nest and 6.5 km Lyrebird circuit trails descend the eastern escarpment to cool temperate rainforest where mosses and tree ferns grow in the damp understorey of giant beech trees. Both tracks include sections through heath and subalpine woodland which flower in late spring.

The Cascades Walk below Wrights Lookout offers yet another look at beech forest before continuing down the valley slope to the cascades, pools and subtropical rainforest at Five Day Creek. This 7 km circuit returns to the tableland through open woodland, making it a particularly appealing walk for birdwatchers, who should see many interesting species along the way.

Dense forest makes off-track walking below the escarpment extremely difficult even for experienced bushwalkers; however, fire trails and unmarked ridge routes provide some exciting overnight bushwalks into the Bellinger River valley.

A scene in New England National Park.

GUY FAWKES RIVER NATIONAL PARK

The tablelands of New England extend eastwards into beautiful river country at Guy Fawkes River NP. This mostly undeveloped park is a little off the beaten track, and often bypassed in favour of nearby rainforest parks.

Roads and woodland trails along the escarpment edge lead to lookouts over a broad, forested valley. The river and its tributaries meander through sandy reaches 600 m below sandstone cliffs, over which spill waterfalls, and steep slopes of rubbly scree.

A rough walking track at Chaelundi Rest Area follows a narrow spur down to Guy Fawkes River. Bushwalkers can continue up river to Jordans Trail which climbs up the escarpment to Spring Gully picnic area. As a 2 day, 30 km circuit this walk allows plenty of time to enjoy the river and surrounding rainforest.

Ebor Falls, on the Guy Fawkes River.

NEW SOUTH WALES

ACCESS

100 km NE of Armidale, 60 km W of Dorrigo. Conventional vehicle access to Chaelundi Rd on eastern boundary or from Hernani and Dundurrabin.

CAMPING

Basic facilities at Chaelundi. Conditions apply to camping on overnight bushwalks.

WHEN TO VISIT

Any season, but bushwalking conditions are best during spring and early summer.

SAFETY

Be careful of submerged objects when swimming or canoeing. Scree slopes are slippery underfoot.

Guy Fawkes River NP is in one of the loveliest areas of the Great Dividing Range.

ACCESS

Visitor centre is 33 km W of Coonabarabran via the John Renshaw Pkwy, 80 km NE of Gilgandra via the Oxley Hwy and Tooraweenah.

CAMPING

Fees charged at all sites. Good facilities and powered sites at Camp Blackman. Basic facilities at Camps Wambelong, Elongery and Pincham. Large groups need to book. Permit required for overnight bush camping at designated sites. Bookings required for Balor Hut.

WHEN TO VISIT

Walking is comfortable from April through October with wildflower displays starting mid-August.

SAFETY

Carry drinking water when walking. Roads can be dangerous in wet weather.

A blade of ancient volcanic rock.

This view of the Warrumbungle Range features the spectacular formation known as the Breadknife.

The Warrumbungle Range looms above the western plains like the ruins of a fortified castle. Its narrow ramparts, delicate spires and heavy domes are buttressed by wooded spurs and valleys. This dramatic landscape is the eroded remains of an ancient shield volcano on the western slopes of the Great Dividing Range.

Known as the land where east meets west, the Warrumbungles are a biological change-over zone. Open eucalypt woodlands cover most of the park, however, its cool, humid gullies shelter hardy rainforest plants, and aridland grasses and shrubs occur on exposed, rocky slopes.

A short walk to Whitegum Lookout near the eastern park entrance sets the scene for a Warrumbungle adventure. The 500 m woodland trail leads to tantalising views of the range and its volcanic peaks. The track has information signs and seats along the way, and a sealed surface suitable for wheelchairs.

Further down the valley, an extensive track system fans out from centrally located visitor facilities. The walks vary from easy 1 km creekside strolls to the challenging Grand High Tops and Mt Exmouth treks, which are between 12 and 18 km. Some of the climbs are steep – Mt Exmouth is the highest peak in the park – but the 360° views are breathtaking.

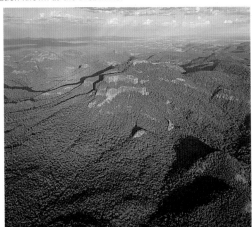

The Warrumbungles are the remains of an ancient volcano.

Siding Spring Observatory is just outside the park.

The shorter, easy-to-medium grade creek walks include sandstone and volcanic rock formations and excellent views to the mountains and valley flats. These walks also take in three main vegetation types with an opportunity to do some wildlife spotting.

Emus, Red-necked Wallabies and Eastern Grey Kangaroos frequent the grassy flats, while goannas, Koalas and wallaroos prefer the woodlands. Forests fringing the creeks attract Swamp Wallabies. Many birds, including parrots, fairy-wrens, robins and honeyeaters, are found in all habitats. Watch for Platypus, water dragons and tortoises in the creeks.

The grand beauty of the mountain summits has beckoned rock climbers and bushwalkers for over 60 years. With its all-encompassing views and spectacular volcanic rocks, the Grand High Tops circuit has become a classic Australian nature walk.

The going is not especially difficult on the well-maintained track, but it's a full day effort with some strenuous climbs. As the trail rises along Spirey Creek, cypress and grass trees join the woodland gum trees. Heath plants take over as the slopes become steeper until finally only the hardiest of stunted shrubs survive in crevices on the barren rock faces.

Eastern Grey Kangaroos can be seen on grassy flats.

Tucked away in this semi-arid land is a waterfall grove of rainforest ferns and Port Jackson figs at Hurleys Camp. Those with bush camping equipment can spend a more leisurely 2 or 3 days exploring all the summit side tracks and the Western High Tops track to Mt Exmouth. Striking sunsets and dazzling night skies are a bonus for the overnight bushwalker. In springtime, there are wildflowers throughout the park and birdwatchers will find many species nesting. Rock-climbing is permitted in most areas, but permits must be obtained beforehand.

Walking trails lead to many scenic lookouts throughout the Warrumbungles.

Flocks of Galahs enliven the edges of the bushland.

Warrumbungle NP is noted for striking sunsets and, later, dazzling displays of stars.

ACCESS

770 km NE of Sydney.
Flights from Sydney and Brisbane.
Visitor numbers restricted.

CAMPING

Guest lodges and self-contained apartments. No camping.

WHEN TO VISIT

November through May, although it's best to avoid the December–January holiday season crowds.

The Lord Howe Island Group was placed on the World Heritage List in 1982 and is included here because of that status.

Red-tailed Tropicbird and chick.

Looking across Lord Howe Island to Mt Lidgbird and Mt Gower.

Lord Howe, with its reef-enclosed lagoon and rainforested mountains, is the ultimate New South Wales nature escape.

There are no glitzy high-rise developments on this verdant crescent of volcanic rock and coral sand, which, with its neighbouring islands, is the eroded remnants of a massive shield volcano that rose from the sea some 7 million years ago. The world's southernmost coral reefs fringe these volcanic outcrops, creating a mid-Tasman transition zone for tropical and temperate marine plants and animals.

The island also has a number of native bird species, including the endangered Lord Howe Woodhen, now being bred in captivity. The area is also important as a roosting, feeding, mating and nesting location for many thousands of seabirds. Here, too, there is an overlap of warm and cold climate species, such as the Red-tailed Tropicbird and Providence Petrel.

The area's reefs and waves attract divers, sailors and surfers.

Balls Pyramid rises over 500 m from the ocean.

More than 240 type of plants have been found among Lord Howe's palm forests, woodlands, highland and lowland rainforests and grasslands. About one-third of these occur nowhere else.

There is no evidence of human habitation prior to the first recorded sighting of the island in 1788. It was first settled in 1833 and the early settlers traded livestock and small crops with passing ships. The trade in Kentia palm seeds now supplements the island's tourism industry.

Summer is the peak visitor season with all facilities booked out months in advance. Diving and boating enthusiasts may also want to avoid the uncomfortable winter westerlies.

There are several good dive spots in the lagoon which varies in depth from 1 to 3 m, and the outer reef edges are about 1 km from the western shoreline. Visitors can enjoy underwater viewing from glass-bottomed boats, deepwater fishing trips and sightseeing excursions to Balls Pyramid and adjacent islands.

Cycling and walking are the most common modes of travel on the 11 km long island. There are numerous trails through the interior to sandy beaches, rocky headlands, seacliffs and intertidal platforms. Adventurous bushwalkers can take a guided trip through creek gullies and rainforest to the 875 m summit of Mt Gower.

On Lord Howe, it is never far to ocean or mountain.

Driving through the island's forest.

A dinghy lies waiting on a Lord Howe beach.

Feeding fish at Neds Beach.

Lord Howe Island was forced above the sea by volcanic activity around 7 million years ago.

ACCESS

32 km S of Sydney or 47 km N of Wollongong via Princes Hwy.
Train and bus services.
Ferry from Cronulla.

CAMPING

Fees charged and bookings required for camping and caravan sites at Bonnie Vale. Permit required and conditions apply to bush camping.

WHEN TO VISIT

Any time of year. Weekends and holiday periods are very busy.

SAFETY

Watch for changing tides and weather conditions on coastal walks.

The popular Eastern Rosella.

Audley offers picnic areas and the opportunity to hire canoes and boats in which to explore the Hacking River.

Australia's first national park was established in 1879, when an area of 7000 ha south of the Hacking River was set aside for use by the people of Sydney.

The reserve was managed in line with what were then common notions of what parkland should be. Riverside forest was cleared for ornamental gardens, exotic trees and animals were introduced, a holiday village and guest houses were built and roads criss-crossed the entire park. Grazing, mining and timber leases were made available. During both world wars, large areas were cleared for military purposes. In the late 1960s, environmental conservation gained a major role in the park's management and today Royal NP covers 15 069 ha and continues to offer city dwellers a coastal bushland retreat.

Regenerating heathland covers most of this low, sandstone-capped plateau, while inland ridges carry woodlands of she-oak, wattle and eucalypts. Remnant stands of subtropical rainforest linger in damp valleys and along gorges cut by the Hacking and its tributaries.

To the east, wind and wave action have created a 30 km coastline of honeycombed and undercut cliffs interspersed with fine sandy beaches and coves. Sheltered gullies behind some of the bays harbour rainforest plants.

Royal NP is well serviced with sealed roads and a 150 km walking track system. There is good road access to inlets along Port Hacking's southern shores and, at the historic Audley village, canoes and dinghies can be hired to explore the river and its tributaries. Audley is also a starting point for some full-day walks, including a riverside valley walk on what used to be Lady Carrington Drive and an 11 km ridgetop walk that drops through rainforest to Uloola Falls on the way to Waterfall township.

The 26 km Coast Track between Bundeena and Otford is one of the park's most popular long-distance walks. The 2-day trek takes in sculpted headlands, beaches, flowering heathlands and rainforest gullies. Many of these features can be reached on shorter walks from Bundeena Drive or on trails leading from carparks behind the surf beaches at Wattamolla and Garie.

A peaceful stretch of water at Wattamolla Beach.

Belmore Falls tumbles over two sandstone ledges.

It has taken millions of years to create the wild and beautiful landscapes of Morton NP. Sediments eroded from a youthful Great Dividing Range were compacted into a basin beneath vast coastal swamps. The transformed rock was then uplifted into a high sandstone plateau which the Shoalhaven and Clyde Rivers have carved into cliff-rimmed gorges.

A patchwork of open forest, woodland, heath, and sedge swamp covers the flat-topped tableland. Below the cliffs, tall eucalypt forest grows on the slopes and damp gullies hide pockets of cool ferny rainforest. On the deep soils of the eastern valley floors red cedar mingle with sassafras and coachwood above lilly pillies, lawyer vine, tree ferns and stinging trees. In the undeveloped southern region of the park, flat-topped monoliths stand above horizontal cliffs before the land folds into the steep ridges of the Budawang Range. There are excellent views of the Budawangs from Pigeon House Mountain, west of Ulladulla. The 3 km, ladder-assisted climb to the summit of this volcanic plug also offers great coastal views.

The spectacularly eroded sandstones of the north Budawang Range provide some of the most exciting bushwalking in NSW. An unmarked but well-worn trail from the Wog Wog rest area begins a 3- to 5-day return walk through Monolith Valley, with its fascinating sandstone formations, to the Castle, from which the views are superb but the climb is difficult.

Morton's northern waterfalls and escarpments have been attracting sightseers since the late 1800s. Easy grade trails along the clifftops at Fitzroy and Belmore Falls offer striking views of Shoalhaven tributaries dropping over layers of sandstone blocks into narrow gorges. A little further west, at Gambells Rest, a network of roads and walking tracks lead to waterfalls, fern gullies and cliffside lookouts over Bundanoon Valley.

The roadless and untouched Ettrema wilderness area, where sheer-walled gorges have almost separated a fork-like plateau from the rest of the highlands, lies south of the Shoalhaven River. Seasoned bushwalkers can access the top of the plateau from a road west of Sassafras or negotiate boulder-lined creeks into densely forested gorges from Yalwal.

The view from Hindmarsh Lookout.

Glorious Fitzroy Falls plunges over a sheer cliff.

ACCESS

155 km SW of Sydney or 150 km NE of Canberra. Northern section from Nowra, Kangaroo Valley, Moss Vale and Bundanoon. Central and southern areas from Braidwood, and several roads off the Princes Hwy between Nowra and Batemans Bay.

CAMPING

Fees charged and bookings required at Gambells Rest. Conditions apply to bush camping.

WHEN TO VISIT

Any time of year. Heath and rainforest flowers from August through November.

SAFETY

Advise park staff of itinerary and carry drinking water in summer if bushwalking in remote areas.

Tianjara Falls.

ACCESS

28 km SE of Nowra via the Princes Hwy and Jervis Bay Rd.

CAMPING

Bookings required and fees charged. Good facilities with unpowered tent and caravan sites at Green Patch. Basic facilities for tents only at Bristol Point and Cave Beach. Bush camping not allowed.

WHEN TO VISIT

Any time of year. Heath wildflowers late August through October.

SAFETY

Take care when walking near cliff edges or on exposed rocky shorelines.

Booderee NP juts out from the Illawarra coastline forming a protective arm around the crystalline waters of southern Jervis Bay. Battered seacliffs, rocky platforms, tranquil coves and pristine beaches line the peninsula's shores.

Jervis Bay, known as Booderee, or "bay of plenty", to the local Aboriginal people, is a popular diving and boating location. Rocky reefs, sand and seagrass beds carry a wealth of marine life. Plants and animals crowd together in these cool southern waters covering every available surface, filling each nook and cranny.

The dramatic cliffs of Point Perpendicular.

The southern headland of Jervis Bay is part of the park.

At low tide, rockpools and crevices reveal the secret life of crustaceans, molluscs, sea stars and anemones. Their fossilised ancestors can be seen in the layered sandstone cliffs. Migrating whales pass by the enclosed bay, but penguins, fur-seals and dolphins frequent its calm clear waters.

Booderee's bounty extends inland through estuarine and freshwater wetlands, coastal heaths, woodlands and remnant rainforest. Of the abundant wildlife, over 180 kinds of bird find food and shelter within the park, along with many different mammals, including possums, gliders, and kangaroos.

Sea cliffs from Cape St George.

The cobalt blue waters of Jervis Bay.

The park is separated into two sections, well-serviced by roads and walking tracks. In the western section, Wreck Bay Beach stretches down to Sussex Inlet, a narrow, twisting estuary leading to the shallow waters of St Georges Basin. Between the two lies an old dune system which supports banksia woodland, eucalypt forest and some interesting freshwater lakes.

Booderee Botanic Gardens, just off Cave Beach Road, were set up in 1951 as an offshoot of the Australian National Botanic Gardens. It provides for the study and propagation of native coastal plants. The 80 ha gardens surround Lake McKenzie with a mix of natural bushland and cultivated native plants. Short walks through the gardens feature information displays, labelled plants, a rainforest gully and heathland.

The eastern section provides access to Jervis Bay and the rugged coastal cliffs. Lookouts and clifftop circuit trails offer excellent views.

Left to right, top to bottom:
Sea jellies; Truncate Coralfish.
Sea star; sea anemones;
Big-belly Seahorse.
Eastern Blue Devil; cuttle,
Half-banded Seaperch.
Yellow zoanthids.

KOSCIUSZKO NATIONAL PARK

ACCESS

450 km SW of Sydney. 188 km SW of Canberra via Cooma. Car and bus services. Numerous access points. Main road through is Snowy Mts Hwy. Some roads closed in winter. Wheel chains required in winter.

CAMPING

Bookings required for sites and rental cabins at Sawpit Creek. Basic facilities located along Alpine Way, Barry Way and the lower Snowy R. Conditions apply to bush camping.

WHEN TO VISIT

Winter for skiing. Autumn and summer for bushwalking. Wildflowers best January and February.

SAFETY

Be prepared for sudden blizzards any time of year.

Summer frees foothills from snow.

Winter brings snow to the peaks of the Australian Alps.

Kosciuszko is the main link in a chain of alpine national parks and reserves stretching between the ACT and central Victoria. A series of worn granite peaks, several exceeding 2000 m, cap this high plateau of rolling alpine meadows. Long narrow valleys and steep western gorges cut into the plateau carrying rain and melted snow to river systems and reservoirs.

Dense forests with fern gullies and stands of giant alpine ash shelter on the valley floors and lower slopes. With increasing altitude, straight-trunked snow gums of the shrubby woodlands become gnarled, multi-stemmed individuals marking the edges of the alpine plains. In summer, wildflowers carpet the treeless meadows in an intricate pattern of heath, sphagnum bogs, herbfields and grasslands.

Alpine creeks feed some of Australia's major rivers.

Snow gums glow in afternoon light.

From late June through September, the park's main range becomes the winter playground of downhill skiers. But resorts and groomed runs occupy less than 1% of the largest national park in NSW.

The beauty and isolation of the Snowy Mountains' natural snowfields have been attracting cross-country skiers since the 1860s. There are several unmarked but well-known routes over snow-blanketed country between the peaks and glacial lakes. Flagged trails around the resort areas offer good full- and half-day outings or can be used to begin longer excursions through the back country and lower plains.

By November the snow has melted and summer brings ideal conditions for walking on the rooftop of Australia. Numerous alpine tracks of varying length and difficulty feature delicate wildflowers, pretty glacial lakes and stunning views across barren peaks to distant horizons.

The two most popular Main Range trails ascend Mt Kosciuszko via a 24 km ridgetop circuit from Charlottes Pass or a 12 km return walk from the top of the Crackenback chairlift at Thredbo. Shorter trails taking in cascading mountain creeks and lush forest centre on the Sawpit Creek area.

Ski resorts take up only a tiny portion of Kosciuszko NP, but they attract many visitors each winter. The national park includes some popular cross-country trails, including the stretch between Kiandra and Kosciuszko and between Charlottes Pass (above) or Thredbo and Kosciuszko. It is possible to overnight in huts once used by mountain graziers.

There are endless possibilities for long bushwalks. The open countryside makes navigation with a map and compass relatively easy, but backpackers need to be prepared for cold, wet weather even in summer. Main Range offers some exhilarating walking, but many prefer the solitude of the Jagungal, Byadbo and Bogong regions.

Those looking for a journey of epic proportions can tackle all or part of the 655 km Australian Alps Walking Trail which runs from Namadgi NP through Kosciuszko to Baw Baw NP. Kosciuszko is one of the few national parks in Australia with areas where horse riders are welcome. The Bicentennial National Trail picks up some of the old stock routes through the high plains country.

A popular attraction in the northern section of the park is Yarrangobilly Caves, four of which are open to the public. There is a thermal pool at Yarrangobilly, as well as picnic areas and a number of short walks.

The Corroboree Frog lives in moss beds in the Alps.

The Common Wombat grazes on high meadows.

A peaceful winter scene in Kosciuszko NP.

MUNGO NATIONAL PARK

The lunette dunes of Mungo NP were formed from tiny particles of clay blown from drying lakes by the wind.

ACCESS

Unsealed roads via Arumpo 110 km NE of Mildura or 150 km NW of Balranald. Check road conditions after rain. Light aircraft landing strip in park.

CAMPING

Fees charged. Basic facilities at Main Camp and Belah Camp. Bookings required for bunk accommodation.

WHEN TO VISIT

Winter is most comfortable for walking.

SAFETY

Carry extra water, fuel and vehicle parts.

The fossil landscape of Mungo NP tells a remarkable story of Aboriginal life and dramatic climate change spanning 40 000 years. The Willandra Lakes Region, which includes the park, was added to the World Heritage List in 1981 because of the significance of its archeological and physical features.

Until about 15 000 years ago, Lake Mungo was one of several inland lakes fed by the Lachlan River overflow. As the climate became drier and warmer, the freshwater lake gradually dried up to become a dry basin of saltbush and grass. A 30 km dune of sand and clay, known as the Walls of China, now marks the lake's north-eastern shoreline.

Wind and rain have cut into the surface of the fossilised dune, unearthing the middens, stone tools and burial sites of the people who lived in this once verdant land.

The Golden Perch, freshwater shellfish, yabbies and the buffalo-sized mammal, Zygomaturus, are long gone from the landscape, replaced by a variety of animals able to cope with a semi-arid environment. Some of the more commonly seen animals and birds include Red and Western Grey Kangaroo, Emu, Shingleback Lizard, Orange and White-fronted Chats.

Trails and a signposted 60 km circuit drive guide visitors through times past and present at this archaeologically significant park on the western plains of NSW.

Shingleback Lizard.

Educational material in the park.

Mungo's distinctive sand dunes conceal relics of a past civilisation, including the world's oldest human cremation.

Mootwingee NP rises above the desert plains of north-west NSW through a series of stark ridges and narrow gorges before spreading across an eroded tableland.

For over 10 000 years the sandstone of Bynguano Range has stored the Wilyakali people's record of their lives and spiritual beliefs. Engraved, stencilled and painted images are scattered throughout the gorges, under ledges and on rockfaces. From April to November Aboriginal rangers guide visitors through some of the finest art sites in NSW. A cultural centre is situated about 7 km inside the park.

Most of the park is designated wilderness; however, walking trails and scenic drives provide an opportunity for self-guided tours of the western ridges and gorges. A 3 km walk into Mootwingee Gorge ends at a delightful rockpool beneath imposing mineral-stained cliffs. Rock art, unusual sandstone formations and ridgetop views can be found along the linked trails at Homestead Gorge. There is also a 3.5 km ridge circuit for sunset views over the Bynguano Range and desert plains. Some of the park's European history, which includes links with Burke and Wills, is revealed along the Old Coach Road drive to Gnalta Lookout.

Most of Mootwingee is designated wilderness.

A colourful cliff-face in Mootwingee NP.

ACCESS

130 km NE of Broken Hill. Unsealed roads impassable after rain. Light aircraft landing strip in park.

CAMPING

Basic facilities at Homestead Creek. Bookings required and fees charged. Permit required for camping in wilderness area.

WHEN TO VISIT

April to November for comfortable walking and guided art site tours.

SAFETY

Carry extra supplies including drinking water, fuel and vehicle parts. Take water when walking.

A Common Wallaroo.

47

STURT NATIONAL PARK

ACCESS

Conventional vehicle access 330 km N of Broken Hill via Tibooburra, 430 km W of Bourke via Wanaaring. Both may be impassable after rain.

CAMPING

Fees. Basic facilities at Dead Horse Gully, Mt Wood, Olive Downs, and Fort Grey. Booking for bunks at Olive Downs. Bush camping not recommended but permits available.

WHEN TO VISIT

Winter and spring. Summer 50°C+.

SAFETY

Carry extras, especially water. Stay with vehicle if breakdown occurs.

The north-west sandy region after rain.

Stony gibber plains roll through the eastern section of Sturt NP to the jump-ups (escarpments) and gullies of Grey Range. Beyond the 150-m high mesas, rows of red sand dunes rise into the Strzelecki Desert.

Rain brings unexpected life to this semi-arid landscape. Colourful wildflowers carpet the plains and waterbirds appear on the billabongs. Major access roads lead to all sections of the park. In the east, scenic loop roads and short walking trails focus on the Jump Up country. The western road traverses clay pans and sandhills to Fort Grey where Charles Sturt established a depot for his 1844 inland sea expedition.

Red Kangaroos can be seen near roads in the park.

KINCHEGA NATIONAL PARK

ACCESS

110 km SE of Broken Hill on sealed roads via Menindee.

CAMPING

Fees charged. Basic facilities at Lake Cawndilla and numerous riverside sites. Bookings required for bunks at shearers quarters.

WHEN TO VISIT

Winter and early spring. Wildflowers cover the plains after rain.

SAFETY

Carry drinking water. Gravel roads in the park — check road conditions.

A road through Kinchega NP.

The lakes of Kinchega NP are part of a larger system carrying the overflow of the Darling River as it meanders across the western plains.

The area supported a large Aboriginal population 15 000 years ago. By the 1870s, the Paakantji people's lifestyle had been overtaken by European grazing interests. Menindee and Cawndilla Lakes are now a haven for breeding waterbirds, while Emus and kangaroos roam the sparsely wooded sandhills and grassy plains. Walking trails and scenic drives follow the Darling River and lake shores, highlighting the park's natural and human history.

The Darling River flows through Kinchega NP.

VICTORIA

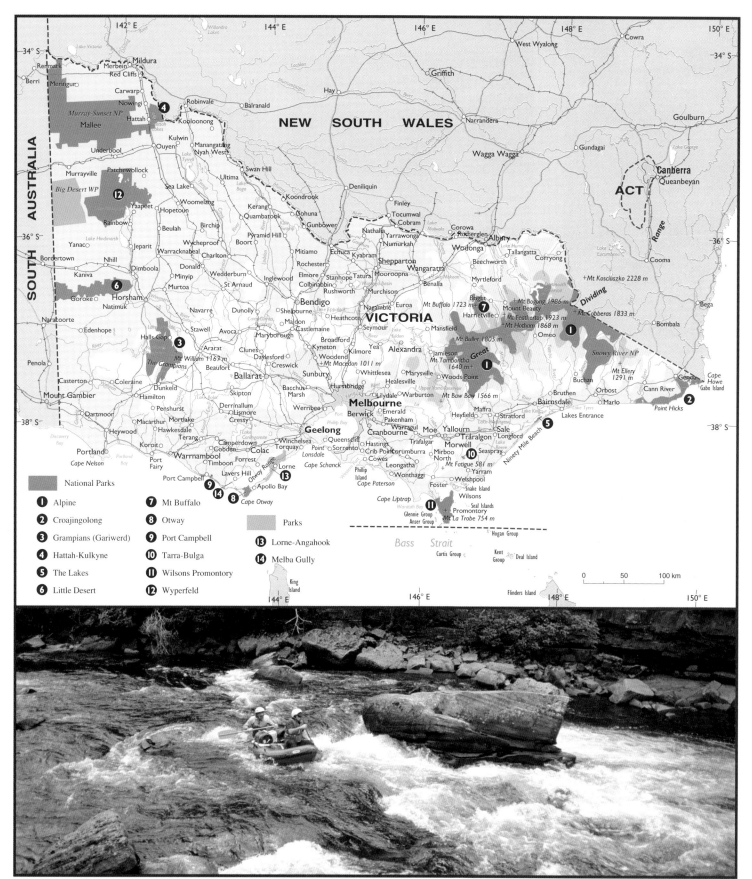

Many of the rivers that have their headwaters in Alpine NP provide exhilarating whitewater rafting adventures.

National Parks

1 Alpine
2 Croajingolong
3 Grampians (Gariwerd)
4 Hattah-Kulkyne
5 The Lakes
6 Little Desert
7 Mt Buffalo
8 Otway
9 Port Campbell
10 Tarra-Bulga
11 Wilsons Promontory
12 Wyperfeld

Parks

13 Lorne-Angahook
14 Melba Gully

WILSONS PROMONTORY NATIONAL PARK

ACCESS

200 km SE of Melbourne via Meeniyan on the Sth Gippsland Hwy.

CAMPING

Fees charged for cabins and unpowered camp sites at Tidal River. Ballot system for Christmas holidays, drawn July. Booking required during holiday periods. Fees and conditions apply to designated bush camping sites; booking and permit required. Privately-owned powered camp sites at Yanakie.

WHEN TO VISIT

Spring for wildflowers, summer for beach activities, autumn for comfortable bushwalking.

SAFETY

Sudden storms make coastal waters dangerous.

Australian Fur-seals underwater.

The rugged granite coast at Wilsons Promontory.

A view from Mt Oberon over Tidal River village.

Wilsons Promontory is one of Victoria's oldest and most popular national parks. This rugged mass of granite juts 40 km into Bass Strait, joined to the mainland by a narrow isthmus.

Between the prominent headlands lie sheltered coves and long beaches. Forested gullies and windswept ridges rise from the coast to central granite peaks reaching 754 m at Mt Latrobe. The Brataualung people have called this place Yiruk since the time it formed part of a land bridge to Tasmania. These days the peninsula is commonly known as the Prom. Eucalypt woodlands, with an understorey of grass, shrubs and gnarled banksias, cover most of the granite mass. Exposed ridges and headlands carry narrow-leaved heath plants. On the higher slopes, Crimson Rosellas *(left)* fly through stands of tall mountain ash and remnant rainforests of sassafras and myrtle beech grow in cool, damp pockets. Warm temperate rainforests of lilly pilly, tree ferns and epiphytes can be found in the lower creek gullies, where paperbarks border peat-brown streams.

A lighthouse stands upon the magnificent granite dome of South East Point.

The isthmus dunefields of she-oak and tea tree scrub flatten out through grassland, heath and paperbark swamps to the mangroves and mudflats of Corner Inlet.

The Prom's surrounding marine parks protect a diversity of underwater habitats ranging from seagrass and sand beds to kelp forests and rocky reefs. Prime cold water dive sites around the coast feature sheer drop-offs, shipwrecks, caves and pinnacles.

There is a wide variety of habitats on and around Wilsons Promontory, and wildlife is abundant. About 90 of the 230 recorded bird species breed there, and fur-seals, wombats, grey kangaroos and Emus are some of the larger animals seen most often.

Whale Rock, a landmark at Tidal River.

One of the many picturesque spots to be seen on the Lilly Pilly Gully walk.

A 30 km scenic drive between the park entrance and the Tidal River visitor centre provide easy access to west coast headlands and beaches. A variety of short nature walks and longer trails across the peninsula branch off the main road, forming an 80 km walking track system, passing scenic areas that were once sites of commercial logging, sealing and whaling.

Self-guided nature trails at Millers Landing, Squeaky Beach, Lilly Pilly Gully and Mt Oberon are excellent introductions to the Prom's main habitats. The popular 2–3 day, 35.5 km southern range circuit features magnificent views, hidden coves and the mainland's southernmost lighthouse at South East Point. Another long circuit of the northern region traverses a greater range habitats, making it an ideal wildlife walk.

The views to Bass Strait from the headlands and hilltops of Wilsons Promontory NP are spectacular.

ACCESS

Sealed road to Mallacoota or gravel road from several points on Princes Hwy from NSW border to Cann River. Check road conditions after rain. No road access to Sandpatch and Cape Howe Wilderness Areas.

CAMPING

Basic facilities at Shipwreck Creek, Wingan Inlet, Mueller River, Thurra River, Peach Tree and Tamboon Inlet. Fees charged, booking required in peak holiday periods. Bush camping permitted on overnight walks.

WHEN TO VISIT

Spring through autumn.

SAFETY

Watch for changing tides and weather conditions on coastal walks.

An obelisk and lighthouse stand on Point Hicks, named for the young *Endeavour* officer who sighted the promontory in 1770.

Croajingolong is one of the finest coastal parks in south-eastern Australia. It is a place of outstanding beauty where mountain-fed rivers run through unspoilt forest valleys and designated wilderness areas to a superb coastline featuring tranquil inlets, immense dunes, long beaches and rocky headlands.

This 87 500 ha park is also a biological treasure house with a thousand species of native plants creating diverse habitats ranging from wind-shorn heaths to hidden rainforest gullies.

Croajingolong's shorelines, heathlands and forests support over 300 kinds of birds, including Satin Bowerbirds, Glossy Black-Cockatoos, Eastern Bristlebirds and the rare Ground Parrot. Mammals and reptiles are also abundant but seldom seen, apart from scavenging goannas.

Gravel roads linked by 4WD tracks provide access to the main waterways and camping areas where canoeing and walking are the best options for exploring this World Biosphere Reserve.

Three short nature walks introduce visitors to the park's main habitats. The Double Creek loop trail, just outside Mallacoota, takes in warm temperate rainforest and eucalypt woodland. Wetlands and heath feature on the Wingan Nature Walk, while the Dunes Walk from the Thurra camping area traverses woodlands and expansive sand dunes.

Another three short tracks offer a look at forest gullies, cascades and rockpools along the Wingan and Betka Rivers and at the base of Genoa Falls. The 2 and 8 km return trails to Genoa Peak and Mt Everard are steep and require physical fitness, but the excellent all-round views are worth the effort.

The 35 km coastal trek from Mallacoota to Wingan Inlet is popular with bushwalkers who want to explore the Sandpatch Wilderness Area. The rugged Cape Howe region is another walk-in wilderness area. Its most accessible feature is Lake Barracoota which is slowly being swallowed by 30 m high sand dunes.

Shipwreck Creek, a popular camping and swimming spot.

A secluded beach at Honeymoon Bay.

TARRA-BULGA NATIONAL PARK

Mountain ash and tree ferns in Tarra-Bulga NP.

Tarra-Bulga is a small park tucked away on the southern side of the Strzelecki Ranges. Sheltered slopes and deep valleys carry fine stands of cool temperate rainforest.

More than 30 kinds of ferns grow in the park's damp gullies which channel sparkling mountain creeks over mossy boulders southwards to the coast. Myrtle beech makes its most eastern appearance at Tarra-Bulga, its moss-festooned trunk rising above smaller sassafras along the creeks. On the ridges, towering mountain ash dominate the forest canopy and the beech becomes part of the understorey.

In the Bulga section visitors can take a suspended walkway into the canopy and over a fern-filled gully. The wheelchair-accessible Lyrebird Ridge walk takes a more down to earth look at the leaf-littered forest floor. Visitors may hear, if not see, the Superb Lyrebird, which is an accomplished mimic. There is also a delightful forest walk to Cyathea Falls in the Tarra Valley.

ACCESS

35 km S of Traralgon.

CAMPING

No camping in park. Privately-owned guest house at Balook. Privately-owned caravan parks S of the park on the Tarra Valley Rd.

WHEN TO VISIT

Any time of year.

SAFETY

Can be cold and wet during winter.

THE LAKES NATIONAL PARK

The Lakes NP is a swampy stretch of sand lying in the midst of the Gippsland Lakes. The 2830 ha park, consisting of Sperm Whale Head Peninsula and Rotamah and Little Rotamah Islands, is a wildlife haven.

Scrubby woodlands, swamps, coastal heath and salt marshes create habitats for a variety of mammals, including kangaroos, wallabies, possums, gliders, echidnas, and the rare New Holland Mouse. There are many habitats for birdlife in the park, and well over 190 different bird species. Birds Australia (formerly known as the Royal Australasian Ornithologists Union) runs a bird observatory and nature study courses on Rotamah Island.

Visitors can enjoy the Gippsland Lakes scenery and spring wildflowers from several points along the peninsula loop road. Well-marked walking tracks traverse the park's vegetation communities and there are a number of hides for observing waterbird behaviour.

Footbridges from the island cross Lake Reeve to 90 Mile Beach which separates the lake system from the Tasman Sea.

ACCESS

63 km E of Sale via Loch Sport. Rotamah Island is 5 km by boat from Paynesville.

CAMPING

Emu Bright: fees charged, booking required in holiday periods. Rotamah Island: Group camp sites can be booked; accommodation for birdwatchers at Observatory.

WHEN TO VISIT

Any time of year. Late August through November for wildflowers.

SAFETY

Follow boating safety regulations.

Long strips of sandy beach separate the beautiful lakes of Gippsland from Bass Strait.

ACCESS

Conventional and 4WD access from numerous points including Mansfield, Omeo, Licola and Buchan. Check for winter road closures due to snow.

CAMPING

Basic roadside facilities in the park. Conditions apply to bush camping. Ski resort accommodation nearby.

WHEN TO VISIT

June through September for winter sports. October through April for walking, canoeing, horse trekking.

SAFETY

Be prepared for winter blizzards and sudden summer snowstorms.

A view from Mt Hotham across the Victorian Alps in winter.

Alpine NP covers mountain peaks, alpine plains and wild river valleys in the Victorian high country. Magnificent forests of tall, straight eucalypts rise through the valleys and foothills to sparse woodlands of twisted snow gums edging treeless plains. Alpine NP links with Kosciuszko NP in NSW and Namadgi NP in the ACT – a chain of parks across the rooftop of Australia.

Panoramic vistas, undeveloped wilderness, winter snow and summer wildflowers make the State's largest national park one of its most popular. Touring by car in summer provides a great introduction to this 646 000 ha park.

Sealed roads to the edges of the Wonnangatta, Bogong and Dartmouth regions provide access to short walking tracks, scenic lookouts, placid lakes and marked cross-country ski trails. Beyond the alpine resorts and visitor facilities lie the beautiful and remote Cobberas, Bogong High Plains and Avon Wilderness.

The park offers endless possibilities for escaping the well-worn tourist routes, however, off-track adventures require good maps, local advice and proper equipment.

A 400 km section of the Alpine Walking Trail provides a range of bushwalking adventures from one end of the park to the other. Highlights along the trail include Mt Howitt, Mt Bogong and Mt Feathertop.

Overnight bushwalks and cross-country ski forays at Dinner Plain and along the razorback to Mt Feathertop in the Bogong region are good preparation for more arduous treks to the less accessible Cobberas and Avon Wilderness.

Horse trekkers and cross-country skiers can experience the high plains country on the Bicentennial National Trail which takes in historic mining sites and pioneer stock routes. For experienced canoeists, there are whitewater challenges on the Snowy, Macalister, and Murray Rivers.

Mt Feathertop, in Alpine NP.

The high country of Victoria's Alps in October.

Mt Buffalo, still dusted with the snows of winter.

Mt Buffalo is an imposing mass of granite on the north-western edge of the Victorian Alps, named by explorers Hume and Hovell because of its shape. This bulky outcrop of the Great Dividing Range rises over 1000 m from broad river valleys to a high plateau studded with rock formations. Bare granite forms towering bluffs and sheer walls along the tabletop edges. The park was first reserved in 1898.

Across the plateau, hardy shrubs, tussocky grasses, summer wildflowers and bog mosses grow among subalpine woodlands. Stately forests of Alpine Ash, Mountain Gum and Candlebark clothe the deep gullies and mountain slopes. Wombats, marsupial mice and mountain dragons claim the highlands while possums, Feathertail Gliders, echidnas and lyrebirds inhabit the forests.

Mt Buffalo is the State's oldest ski resort. The historic chalet was built in 1910. It has a good family atmosphere, and its slopes and cross-country trails are ideal for novice skiers.

A 40 km sealed road runs through the park to the base of the Horn, a 1723 m rock pyramid which looms above the plateau. Ninety kilometres of trails branch off the road, offering a variety of half- and full-day walks once the snow has melted.

Short, self-guided nature trails to the Gorge, Dickson Falls and View Point have leaflets introducing visitors to the geological and natural history of the park. The 2 km gorge rim walk features the dramatic Eurobin Falls and near vertical walls popular with rock climbers. The View Point and gorge rim walks lead to excellent lookouts across the plateau to Alpine NP.

Other tracks lead to impressive granite tors and boulder formations such as the Monolith and the Cathedral. Visitors can also enjoy swimming, fishing and canoeing at Lake Catani and park rivers and creeks.

ACCESS

320 km NE of Melbourne. Sealed road off Ovens Hwy at Porepunkah.

CAMPING

Good facilities at Lake Catani. Booking required and fees charged. Open November to May. Lodge and motel accommodation available in the park.

WHEN TO VISIT

Winter skiing. Good bushwalking from November through April.

An icy mountain creek.

A granite bluff overlooks surrounding foothills.

OTWAY NATIONAL PARK

ACCESS

Road access from the Great Ocean Rd between Apollo Bay and Lavers Hill. Some roads may be closed during wet weather.

CAMPING

Basic facilities at Blanket Bay, Aire River and Johanna. Booking advisable during peak times.

WHEN TO VISIT

Spring through autumn. Winters are usually wet and cold.

SAFETY

Be aware of changing tides and weather conditions on the coast.

Otway NP boasts 60 km of rugged coastal scenery against a backdrop of richly forested mountain slopes, tall eucalypt forests and coastal heathlands.

The Otway Range comes to an abrupt end at Bass Strait in the eastern section of the park. Its high rainfall and fertile soils feed dense forests of giant eucalypts. Runoff collects in clear streams which tumble to sandy coves through gullies of ancient rainforest trees and luxuriant ferns.

Jumbled rocky shores and wave-eroded platforms separate the tiny east coast bays in some of the most rugged and inaccessible country in Australia. West of the cape, the park sweeps along the edge of coastal plains to the limestone cliffs of Port Campbell NP.

Beauchamp Falls, a gem of the Otway Range.

Hopetoun Falls are just outside the national park.

A fern gully in Otway NP.

Tree ferns and creek, Melba Gully State Park.

There are several turnoffs from the Great Ocean Road which follow gravel roads through the park to forest and seaside walking tracks. The rainforest walk at Maits Rest is a delightful 800 m trail. Many kinds of trees and ground ferns grow in this cool, damp gully where mosses, lichens and balls of orange fungi clothe the heavily buttressed beech trees.

A 4 km circuit track from the Shelly Beach carpark runs down to the Elliot River mouth, then loops back along the western valley ridge. A fair weather, low tide alternative is to head south along the coast to Geary River where a side track returns to the main route.

Visitors can either drive through the ranges to Cape Otway or include the lighthouse in a full-day circuit walk from the Aire River camping area. Further west, Johanna offers good surf and beach walks.

The Otway forests are home to Yellow-bellied Gliders, the Platypus, wallabies, potoroo and the very rare Spotted-tailed Quoll. Satin Bowerbirds, Rufous Bristle-birds and Grey Goshawks may be seen in the forests, while migratory albatross, shearwaters and waders appear on the coast each spring.

While visiting this area, it is worth exploring Angahook-Lorne State Park, which surrounds the town of Lorne. Erskine Falls is north-west of Lorne and the reserve has many walking tracks. Nearby Melba Gully State Forest, the "jewel of the Otways", is only 73 ha, but is noted for its rainforest remnants, tree-ferns and glow-worms glimmering on the tracks at night.

WILDLIFE OF THE OTWAY RANGES

Some creatures found in the Otway Ranges. *Left and clockwise*: Crimson Rosella; Eastern Yellow Robin; Yellow-bellied Glider; Spotted-tailed Quoll.

Erskine Falls, in Angahook-Lorne State Park, is easily accessible from the town of Lorne.

ACCESS

Several turnoffs from the Great Ocean Rd between Princetown and Peterborough. Visitor Centre at Port Campbell outside the park.

CAMPING

Booking required and fees charged for caravan and camp sites outside the park at Port Campbell.

WHEN TO VISIT

Spring through summer for wildflowers and walking. Winter can be cold and wet but the waves are spectacular.

SAFETY

Beware of changing tides and weather conditions if descending the cliffs. Swimming is not recommended.

Four views of the Twelve Apostles, limestone stacks cut from the mainland by the surging waves of the Southern Ocean.

Port Campbell NP is a narrow coastal strip of limestone cliffs and stunted heaths. Massive ocean waves harried by relentless southern winds have sculpted a haunting landscape from the seacliffs.

The arched headlands, long canyons, blowholes and isolated towers are the work of dynamic forces which continually seek out weakness in the layered limestone.

This beautiful and treacherous coastline has claimed its share of wrecked ships, including the famous *Loch Ard*.

There are spectacular views from numerous clifftop lookouts. Sunsets are superb, especially in summer when flocks of shearwaters return to their nesting sites on the cliffs and stacks and Little Penguins bring food to their chicks waiting in nest burrows on secluded beaches.

Muttonbird Island is a breeding sanctuary for seabirds.

This intricately indented coastline was once known as "the shipwreck coast".

Steps give access to lookouts and beaches.

Little Penguins nest along this coastline.

Short walking trails highlight many of the park's natural and historical features. At low tide during calm weather, visitors can descend the cliffs to a collapsed sea cave at the Grotto and to small beaches at Gibson Steps and Loch Ard Gorge.

Near the gorge, the Blowhole's spumes of seawater erupt from a 100 m long tunnel beneath the cliffs. When the tunnel eventually collapses, the ocean will have new headlands to carve into arches and rock towers.

A self-guiding headland walk from the visitor centre at Port Campbell includes a safe swimming beach.

THE OCEAN AS SCULPTOR

London Bridge in 1990, when its landward span seemed so secure that visitors to Port Campbell NP happily walked over it to admire the view from the further end of the headland.

Three days after the above photograph was taken, the landward span collapsed, stranding sightseers. This procedure of isolating headlands has formed stacks such as the Twelve Apostles.

Quiet coves lie at the bases of the park's seacliffs.

ACCESS

Various access roads from the Western, Henty and Glenelg Hwys.

CAMPING

Fees charged for basic camping facilities at the many sites in the park. No booking required. Conditions apply to bush camping.

WHEN TO VISIT

Spring and autumn for comfortable walking.

SAFETY

Take care when climbing rocks.

The Balconies, one of the Grampians' many spectacular rock features.

Mackenzie Falls are most impressive after rain has fallen.

The bold profile of the Grampian sandstone ranges falls away to Victoria's north-western plains signalling a sudden end to Australia's Great Dividing Range.

Millions of years ago, compacted sediments were heaved upwards from an ancient sea bed, then tilted, folded and eroded into a series of long sinuous ranges separated by shallow valleys. The four main ranges are Victoria, Serra, Mt William and Mt Difficult.

Sweeping western slopes, craggy eastern rockfaces, waterfalls and and cascading streams create a dramatic landscape which is home to an incredible array of plants and wildlife.

Over 1000 kinds of trees and plants grow in heath, wetlands, forests, open woodlands and tree fern gullies. Twenty of these, including the spectral duck orchid, are found nowhere else. Kangaroos, wallabies, parrots, wombats, Emus and possums are often seen in the park. The Feathertail Glider, Smoky Mouse, Platypus and Fat-tailed Dunnart are the more elusive of the mammal species of the Grampians.

The Wedge-tailed Eagle may be seen over the peaks.

Roads and walking trails are as numerous as the park's natural and cultural features, so there is no difficulty getting around. Several tracks branch off the Grampians Road, leading to spectacular eastern views from Mt William Range and to Major Mitchell Plateau where wildflowers carpet the heathland from late August to November.

The Wonderland circuit and Mt Victory Road, west of Halls Gap, take in the park's well-known lookouts as well as Mackenzie Falls and rock formations such as the Pinnacle, the Balconies and Grand Canyon. There's a wonderful 4 km walk alongside the Mackenzie River to a volcanic rockface at the base of the falls.

Looking across the Grampians to the plains beyond.

Where there are rocks, there are people who climb them, and the Grampians are no exception. However, many of these striking formations are central to the Aboriginal people's creation beliefs. More than two-thirds of Victoria's Aboriginal art sites are found within the ranges. The Bambruk Living Cultural Centre at Halls Gap has displays and demonstrations introducing visitors to Aboriginal customs and lifestyle. The centre also runs guided tours to art sites at the northern end of the park near Mt Stapylton.

There are other rock shelters in the more remote Victoria Range, not far from the Buandik camping area, which feature painted and stencilled figures. This range also offers some excellent overnight walks with good rock climbs at the Chimney Pots and Fortress.

A distinctive rock formation.

The lookout on The Pinnacle.

Eastern Grey Kangaroos are common on creek flats and in eucalypt woodland in the park

WYPERFELD NATIONAL PARK

Emus can be seen throughout the park. The male Emu incubates the eggs and cares for the chicks.

Bark for a shield was taken from this tree by an Aborigine.

Daisies star the sand after rain has fallen.

ACCESS

Conventional vehicle access 25 km N of Yaapeet. 4WD access on Murrayville–Nhill track and Gunners Track off Mallee Hwy. Check road conditions after rain.

CAMPING

Fees charged for basic facilities at Casuarina and Wonga. Conditions apply to overnight bush camping.

WHEN TO VISIT

Spring and autumn for walking. Winter days are pleasant but nights are freezing.

SAFETY

Carry drinking water when travelling in the park.

Peaceful Dove.

It takes time for even the most dedicated park visitor to acquire a taste for the semi-arid mallee country. And there is no better start to a life-long appreciation of the mallee than a spring visit to Wyperfeld NP in north-western Victoria.

The discovery of delicate ground orchids or a chance encounter with the rare mound-building Malleefowl soon dispels first impressions of parched lake beds and drab scrub.

The park's undulating sand plains extend west from a chain of dry lakes to the steep white dunes of the Big Desert Wilderness Park. Grassy woodlands of river red gum and black box fringe the lake beds where Emu and Eastern Grey Kangaroo gather at dawn and dusk. Every 20 years or so, the Wimmera River carries enough water to fill the lakes, attracting huge flocks of breeding waterbirds.

Low ridges of cypress pine occasionally interrupt the vast sandy plains of shrubby small eucalypts, broombrush and heath. This grey-green landscape becomes a mass of colour after winter rains as daisies, myrtle, correa and 50 species of ground orchids come into flower.

This relatively undisturbed mallee is a haven for dry country wildlife. Aside from kangaroos and brushtail possums, the park's mammals tend to be small and secretive. Mitchell's Hopping Mouse and the Silky Mouse shelter in deep burrows during the day and scurry out to feed at night. Colourful parrots, honeyeaters, fairy-wrens and robins are just some of the 200 kinds of birds found in the park.

The emphasis at Wyperfeld is on walking rather than driving and there are some good trails near the camping areas in the eastern section. Two of these are self-guided walks for which leaflets, which explain life in the mallee, are available.

Early morning is the best time for wildlife-spotting or photography along the 10 km Lignum Track or the 6 km Desert Walk Circuit. Sunsets in Wyperfeld are glorious.

LITTLE DESERT NATIONAL PARK

Little Desert NP is a low profile expanse of river floodplains, claypans and sandy plains on the edge of Victoria's mallee country. This misnamed park, neatly hemmed by cleared pastures, is not a desert. Mild temperatures and moderate winter rains support a remarkable number of plant and animal communities.

River red gum and black box woodlands thrive on the black loam soils edging the Wimmera River, while yellow gums claim the heavier clay soils around shallow depressions. Mallee eucalypts and desert banksia along with broombrush, grasstrees and heath form a low cover over the sandplains. The delicate flowers of over 50 varieties of orchids appear on the plains during September and October.

Little Desert is a birdwatcher's paradise. Chats, wrens, babblers, honeyeaters and finches are just some of the park's 230 bird species. Ground, mound and tunnel nesters include Emu, malleefowl, pardalotes and parrots. Rain turns the claypans into billabongs and marshes attracting flocks of itinerant waterbirds, Budgerigars and Cockatiels.

An 800 km grid system of 4WD tracks overlies the park. However, camping areas and walking trails are restricted to the eastern section.

Budgerigars at their nest site in a hollow branch.

Short nature walks provide a good introduction to this fragile ecosystem starting with the Pomponderoo Hill lookout over the mallee. Take a spring wildflower stroll on the Stringybark Nature Walk just off the sealed road south of Nhill, or search for elusive Malleefowl on the 1 km Sanctuary Nature Walk south of Kiata.

Spring, when birds nest and flowers blossom, is an ideal time to explore all or part of the Desert Discovery Walk. There are several access points to this 84 km circuit which features riverside woodlands, water-filled claypans and mallee heath.

ACCESS

Sealed roads from Kaniva, Nhill and Kiata on the Western Hwy. Dirt tracks within the park may close after rain.

CAMPING

Basic facilities at Horseshoe Bend, Ackle Bend and Kiata. Fees charged. Permit required for overnight bush camping on the Desert Discovery Walk.

WHEN TO VISIT

Autumn through spring.

SAFETY

Carry drinking water when walking or driving through the park.

When enough rain falls, this dry claypan in Little Desert NP will fill with water and feeding waterbirds.

Grey Teal arrive after rain falls.

ACCESS

Sealed road access from Hattah on the Calder Hwy. Dirt tracks in the park may be closed after rain.

CAMPING

Basic facilities at Lake Hattah and Lake Mournpoul. Bush camping is permitted.

WHEN TO VISIT

Autumn through spring. Winter rains bring life to the lakes and spring brings wildflowers and nesting birds.

SAFETY

Bring drinking water as park supplies are limited. Avoid boggy ground after rain.

Hattah-Kulkyne NP extends to the floodplains of the Murray River, where dawn brings scenes such as the backwater above.

When the Murray River spills into Chalka Creek, Hattah-Kulkyne becomes an oasis of lakes and flooded plains amidst dry mallee scrubland. Scores of waterbirds appear seeking mates, food and nesting sites.

While the lake system rarely floods, winter rains imbue the mallee with a subtle beauty. Shrubby woodlands of black box trees, wattle and bottlebrush rim the floodplains while mallee eucalypts and cypress pine dominate the rolling sandplains.

This is by no means a pristine landscape, having suffered the effects of overgrazing, rabbit plagues and timber harvesting. Gradual expansion of the park and its designation as a World Biosphere Reserve have aided the slow processes of natural revegetation.

An introductory nature walk near the visitor centre can be followed up with a self-guided scenic drive around Lake Hattah or a long circuit walk of the southern lakes. The well-maintained but unsealed Mournpoul Track skirts several lake beds as it runs north through the park. Alternatively, numerous sandy tracks provide bushwalking and 4WD access to the northern lakes and along Chalka Creek.

After rain the plains are covered with grasses and kangaroos.

The lakes are bordered by river red gums.

TASMANIA

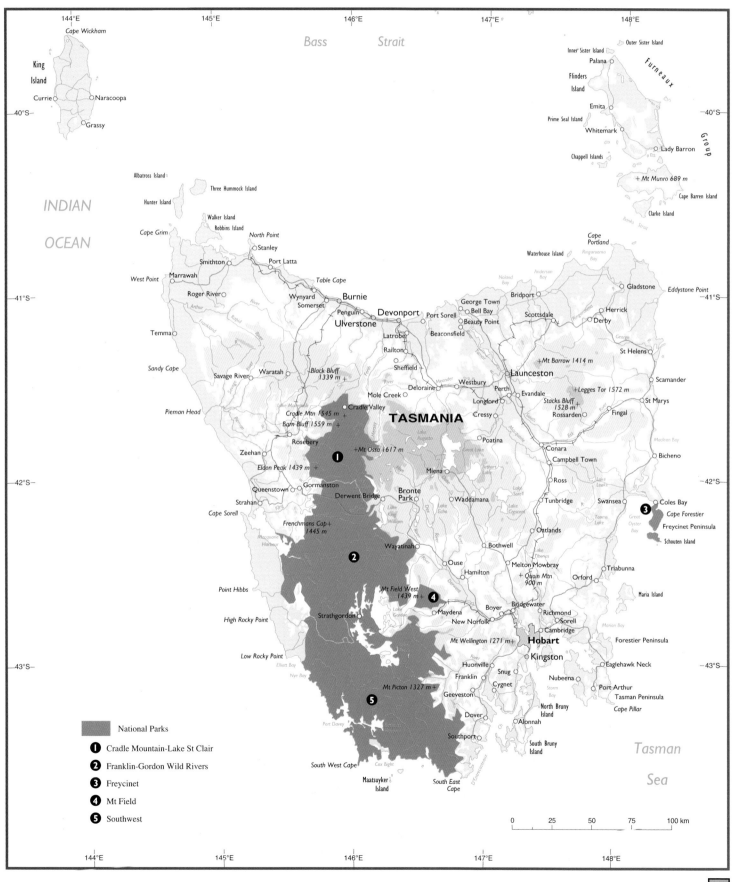

National Parks

1. Cradle Mountain-Lake St Clair
2. Franklin-Gordon Wild Rivers
3. Freycinet
4. Mt Field
5. Southwest

0 25 50 75 100 km

ACCESS

75 km NW of Hobart via New Norfolk and Westerway. 16 km gravel road through park to Lake Dobson. Chains needed in winter.

CAMPING

Fees charged. Good facilities with powered and unpowered sites near park entrance. Booking required for cabin accommodation.

WHEN TO VISIT

Spring through autumn for walking. June to October for winter sports.

SAFETY

Be prepared for cold, wet weather even in summer.

Male Flame Robin at the nest.

In Mt Field NP's cool temperate rainforest.

Mt Field NP, just an hour's drive from Hobart, is the gateway to Tasmania's southwest wilderness.

This 16 265 ha park is one of the few highland areas where the whole family can experience ancient forests, mountain-fed rivers and glaciated landscapes without undertaking long and arduous bush treks.

Visitors enter the park through a lowland valley where Derwent River tributaries cascade through tree fern gullies, and stands of 80 m tall swamp gums dwarf the other forest eucalypts.

It is only a 15 minute walk on a wheelchair-accessible sealed track to see Russell Creek falling 40 m over a series of rocky terraces fringed with dense ferns. The trail continues as a 2 hour circuit to Horseshoe Falls and Lady Barron Falls. Along the way a side trail leads to the Tall Trees loop walk where the ins and outs of life in the wet eucalypt forest are explained.

Russell Falls, one of the most popular attractions of Mt Field NP.

A mountain lake reflects surrounding forest.

Primeval forest. In the foreground is a pandani, a plant whose ancestry can be traced back millions of years.

The Lyrebird Nature Walk, 6 km beyond the park entrance, introduces visitors to the primeval world of Antarctic beech trees found in cool temperate rainforests. Several walking tracks branch off the park road as it climbs to Lake Dobson. Before setting off on these longer walks, it is worth making the steep, but relatively short, climb to Seagers Lookout for panoramic views across the park.

During the ice ages, glaciers advanced and retreated over this high plateau, carving the jagged mountain ranges, scouring the valleys and gouging out lake beds. The glaciers are long gone, but winter snow and ice continue to influence the landscape and highland plant life. The plateau's snow-tolerant pines, heath and alpine herbs differ dramatically from the river valley's lush forests.

The Pandani Grove Nature Walk loops through subalpine vegetation on the edge of Lake Dobson, providing a closer look at unique Pencil Pines and the curious palm-like heath plant known as Pandani.

Well-marked trails across the plateau feature glacial lakes, mountain views and fields of alpine wildflowers. These linked tracks create many walks of varying length and difficulty. Two of the more popular routes are the 5 hour Mt Field East circuit and the 8 hour return walk to the top of Mt Field West, the park's highest summit at over 1430 m.

From July to October, the plateau offers some wonderful cross-country skiing, and natural snow runs at Mt Mawson provide limited downhill skiing.

SOME TASMANIAN MAMMALS

Some Tasmanian mammals. *Above and clockwise:* Eastern Pygmy-possum; Short-beaked Echidna; Tasmanian Devil; Common Brushtail Possum.

A good way to see the calmer stretches of the rivers.

ACCESS

Several access points off the Lyell Hwy between Derwent Bridge and Queenstown. By boat or seaplane from Strachan.

CAMPING

Fees charged for basic facilities at Collingwood R. Riverside bush camping sites for rafting expeditions.

WHEN TO VISIT

Spring through autumn.

SAFETY

Don't underestimate the power of the rivers which rise quickly and unexpectedly.

Franklin-Gordon Wild Rivers NP lies at the centre of the Tasmanian Wilderness World Heritage Area. During the early 1980s it was also at the centre of a bitter conflict over plans to dam the Gordon River for hydro-electricity generation. After 3 years of public protest and political manoeuvring, the controversial proposal was shelved.

The river and its many tributaries continue to run free and wild from rough-hewn mountain ranges through sheer-walled gorges and deep, rainforested valleys.

Near its junction with the Gordon, the Franklin River carves through a belt of limestone where water-eroded caves harbour a rich history of Aboriginal life dating back to the last Ice Age.

On the Franklin River.

A misty morning on the Gordon River, Franklin–Gordon Wild Rivers NP.

The Franklin and Gordon Rivers are considered Australia's most exhilarating whitewater rafting course. The standard run between Collingwood River and Heritage Landing takes about 2 weeks and novice canoeists can make the trip with the help of experienced guides. Those who prefer to keep their feet dry can explore the lower Gordon River on boat or seaplane tours which cross Macquarie Harbour.

Several lookouts and short nature walks along the park's northern boundary branch off the Lyell Highway. Lookouts at King William Saddle, Surprise Valley and Donaghys Hill offer dramatic views over glacier-sculpted valleys and mountains. Easy grade trails at Franklin River, Collingwood River, and Nelson Falls provide an introduction to the rainforest which covers much of the park. High rainfall and melting snow feed these cold climate forests of Antarctic beech, sassafras, leatherwood and native pine.

Visitors can get another look at the Franklin River on a short, 15 minute walk along the start of the Frenchmans Cap Track. The trail continues through button grass peat bogs, forest and woodland to the 1443 m, domed summit of Frenchmans Cap. This is a strenuous 4 to 5 day return walk with some exciting climbing on the cap's vertical quartzite face.

Tall trees tower over one of the park's rivers.

Cool temperate rainforest offers tree trunks decorated with mosses and lichens. *Inset, above:* Platypus.

ACCESS

Lake St Clair is 170 km NW of Hobart via Derwent on the Lyell Hwy. Cradle Valley is about 85 km SW of Devonport via Sheffield or Wilmot. No internal road links. Winter road closures.

CAMPING

Fees apply. Unpowered camp sites at Cradle Mountain; privately-owned lodge accommodation also available. Powered and unpowered sites at Lake St Clair. Bush camping permitted.

WHEN TO VISIT

Bushwalking and alpine wildflowers from December through February. Autumn is less crowded and offers relatively settled weather. In winter, snow, blizzards and cold, clear days.

SAFETY

Weather unpredictable; warm, water-proof clothing essential all year.

Walking the high country.

Heath and button grasses cover the plains between the rugged peaks of Cradle Mountain-Lake St Clair NP.

Water, in all its guises, is the creative element behind the unforgettable, rugged landscapes of Cradle Mountain-Lake St Clair NP.

For over 30 000 years the combined talents of ice, snow, rain and mist have been working their magic from the heights of Mt Ossa to the depths of Lake St Clair. Glacier-scarred mountain ranges bare their volcanic rockfaces above deep river valleys and rolling plains dotted with crystal clear lakes.

Eucalypt forests grow on the slopes below the mountains.

Gnarled gum trees and hardy shrubs survive harsh winds and winter snow on mountainous, rocky slopes. The valleys and sheltered slopes below carry tall eucalypt forests and cool temperate rainforest. On the plains between the valleys and ranges, boggy moors of tea tree, heath, button grass and cushion plants separate ridges of open woodland.

Native pine and beech trees flourish in the park's cool, wet environment. Myrtle beech, Huon and King Billy pine appear as forest giants in sheltered sites, while pencil pines and scrambling beech shrubs known as fagus, or tanglefoot, grow in exposed areas above and below the snow line.

Visitor facilities and walking trails focus on the scenic grandeur of Cradle Valley in the north and Lake St Clair in the south. Family nature walks ranging from half an hour to 2 hours duration feature mountain views, glacier lakes, waterfalls, rainforest and moorlands. Full day walks include summit climbs and lake circuits. The best-known walk in this World Heritage listed park is the Overland Track. Myriad side trails and unpredictable weather make this 80 km north–south traverse a challenging wilderness adventure.

Artists Pool is a place to break a strenuous walk.

The unique vegetation of Tasmania's high country.

Tall pandani grow in the forested area of the park called "The Ballrooms".

Cradle Mountain and Lake Dove form a magnificent tableau.

ACCESS

Cockle Creek is a 2 hour drive west of Hobart via Geeveston and Lune R. Sealed and unsealed access to Lake Pedder from Gordon River Rd about 120 km NW of Hobart via Maydena.

CAMPING

Fees charged. Basic facilities at Huon in the north and Cockle Creek in the south. Bush camping at designated sites on long distance walks.

WHEN TO VISIT

Spring through autumn.

SAFETY

Bushwalkers should lodge itinerary with the ranger.

Southwest NP is one of Australia's true wilderness areas. It is an ancient landscape thrown up from the sea, forged in molten lava and scoured by glaciers. Wind and water now further shape this raw and isolated world on the edge of the great Southern Ocean.

No roads breach the solitude of Southwest NP's chiselled mountain ranges, hanging valleys and glacial lakes. The harsh weather and jagged seacliffs have kept the lowland valleys, deep inlets and scattered beaches from coastal development.

Rainforest, low and dense in the highlands, grows down to the coast, where the trees become taller and shrubs replace the ferns and mosses. Eucalypt forest, impenetrable scrub and button grass plains surround the rainforest. Alpine heath and cushion plants grow above the snowline that is marked by stunted, deciduous beech trees.

Bathurst Harbour is a great place to explore by water.

The rugged and very beautiful coastline of Southwest NP.

South West Cape, Australia's most southerly point.

The south-west's diverse habitats are home to many of Australia's unique animals. Tasmanian Devils, wallabies, quolls and possums are common in the forest fringes while native rats, wombats and rare parrots prefer the open sedgelands. Platypus are abundant, having taken advantage of the park's many creeks, lakes and rivers. And if the Thylacine, or Tasmanian Tiger, still exists, the south-west is its likely refuge.

This vast and remote park is also a retreat for those seeking a demanding wilderness experience. Opportunities for adventure include boating, rock climbing, and long treks. Physical fitness and training in the necessary skills are essential for safety's sake.

The Port Davey Track is a popular 1 to 2 week walk traversing the park from north-east to south-west. The 54 km trail follows river valleys to the shores of Port Davey inlet. Bushwalkers can cross the Bathurst Narrows by boat and continue to Melaleuca to join the Southcoast Track, or catch a light aircraft out of the park. Great scenery, beaches and waterbird wetlands are found along the 80 km coastal trail between Melaleuca and Cockle Creek.

The glacier-carved landscape of the Arthur Range offers some of the park's most rugged bushwalking. The challenging eastern and western range circuits require rock-climbing skills.

A cirque lake high in the Arthur Range, Southwest NP.

This Federation Peak landscape was carved out by ice.

Diving among the giant kelp off Tasmania's seacoast.

This World Heritage listed park is not just the domain of experienced and well-equipped adventurers. There is northern road access to the shores of Lake Pedder and some good, short to mid-length walks, including a 20 minute rainforest nature trail. Further along Scotts Peak Road is a 6 km track to the 1425 m summit of Mt Anne for stunning views over Lake Pedder and across valleys and moulded hills to the serrated peaks of Arthur Range. A 7 km trail offers a challenging walk to Lake Judd, a glacial lake below the cliffs of Mt Anne.

Visitors can also explore the park's wild south coast from Cockle Creek. There is an easy grade track across the headland to South Cape Bay for a 4 hour return walk through forest, heath and sedgelands.

An aerial view of the park's coastline.

ACCESS

180 km SE of Launceston off the Tasman Hwy from Swansea. Alternative access from Bicheno.

CAMPING

Fees and booking required during peak times. Privately-owned facilities at Coles Bay. Basic facilities at Richardsons Beach. Sand Dune and Honeymoon Bay open summer only. Camping at sites with basic facilities.

WHEN TO VISIT

Spring through autumn for bushwalking. Summer for water-based activities.

SAFETY

Carry drinking water on longer walks. Observe fire regulations.

In the distance is the Hazards Range.

With its mild climate and beautiful east coast scenery, Freycinet NP is a relaxing alternative to the rigours of Tasmania's famous wilderness parks. It is an ideal spot for a family holiday.

Freycinet Peninsula and Schouten Island form a granite barrier between the blustering Tasman Sea and the calm shallows of Great Oyster Bay. Lichen-painted rockfaces drop sharply to the open sea while sandy crescents and wooded headlands line sheltered western shores.

Safe swimming beaches, protected boating conditions and coastal scenery are the big attractions at Freycinet NP. There are few roads past the Coles Bay camping area, however, well-maintained tracks and easy terrain offer many hours of pleasant walking through the ranges and along the coast. During summer months, park rangers conduct guided walks and other activities for adults and children.

The Hazards' distinctive northern peaks are separated from the park's central range by an isthmus of sand dunes and wetlands. Schouten Island lies a kilometre off the end of the peninsula. A geological fault splits the island, with Freycinet's red granites reappearing to the west and basalt ridges rising to the east.

The Mt Amos summit walk in the Hazards is a 3 hour return trip involving a fairly strenuous climb, but the views over Wineglass Bay and the peninsula are outstanding.

Classic views of the bay are also a feature of the half-hour lookout walk to the saddle between Mt Amos and Mt Mayson. As well as providing access down to Wineglass Bay, this trail is the first leg of two popular circuit walks. By crossing the isthmus to Hazard Beach then continuing north along the coast, visitors can make a half-day loop around the base of the western Hazards. The alternative is a 2 day, 30 km walk through the central range to Cooks Beach, returning to Coles Bay via the western heathlands and beaches.

A 3 km track from Cooks Beach leads south to Bryan Beach, which stretches between a freshwater lagoon and the deep Schouten Passage. Off-track walks on the tip of the peninsula take in secluded beaches, boulder headlands and sweeping coastal views.

The rugged coastline of Freycinet NP.

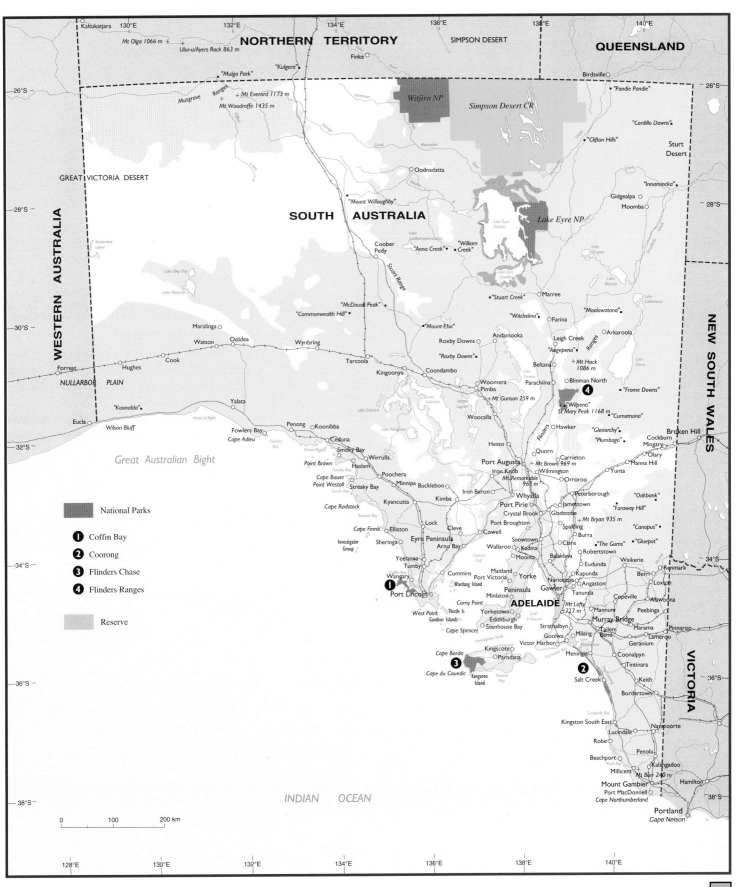

South Australia

Legend:

National Parks

1 Coffin Bay
2 Coorong
3 Flinders Chase
4 Flinders Ranges

Reserve

ACCESS

180 km SE of Adelaide. Numerous access points off the Princes Hwy south of Meningie. 4WD access to Younghusband Peninsula from five crossings south of Salt Creek.

CAMPING

Permit required and fees charged. Basic facilities at Barker Knoll, Salt Creek and 42 Mile Crossing. Bush camping at designated sites on mainland and Younghusband Peninsula.

WHEN TO VISIT

Any time of year.

SAFETY

90 Mile Beach is unsafe for swimming south of Tea Tree Crossing.

Sand dunes on the Younghusband Peninsula.

Coorong NP takes its name from the narrow saltwater lagoon stretching 145 km south from the Murray River mouth behind the high dunes of Younghusband Peninsula. Where the peninsula joins the mainland, the shallow waterways of Coorong Lagoon give way to a series of salt lakes and claypans between the old dunes of previous coastlines.

The Coorong is famous for the number and variety of resident and migratory birds which thrive in its diverse wetland habitats. Australian Pelicans, Silver Gulls and Crested Terns numbering in the thousands have established breeding colonies on the lagoon's island sanctuaries. Permanent residents also include egrets, Black Swans, Black-winged Stilts and many species of duck.

Spring brings migratory wading birds from the northern hemisphere, such as godwits, snipes and stints. About the same time, honeyeaters return to the Coorong's mallee and coastal heath from their winter sojourn in Queensland and New South Wales. The departure of these northern visitors during March and April sees the arrival of endangered Orange-bellied Parrots and their young escaping the harsh Tasmanian winter.

The Coorong is famous for its flocks of Australian Pelicans.

Boating and 4WD are the most popular ways of getting around the park, however, several vantage points just off the Princes Highway offer views across Coorong Lagoon to the haunting dunescapes of Younghusband Peninsula. A 10 minute walk from the Jacks Point carpark leads through inland sand dunes to lookouts over the pelican rookeries. Further south at Salt Creek the 2 km Lakes Nature Trail shows some of the park's plant communities as it crosses low-lying dunes and follows the shores of fresh and saltwater lakes.

There are a number of boat-launching ramps off the Princes Highway between Meningie and Salt Creek. Prevailing southwesterly winds tend to knock up a short, uncomfortable chop averaging 1 to 2 m high on the waters of the lagoon. Parnka Point provides the shortest boat crossing to Younghusband Peninsula.

The peninsula is little more than 2 km at its widest point so it is about a 30 minute walk across the dunes to 90 Mile Beach from the lagoon's eastern shoreline.

42 Mile Crossing is the only crossing south of Salt Creek to provide all year, all weather, 4WD access to the peninsula. The beach north of Tea Tree Crossing is closed to vehicles during the Hooded Plover breeding season from late October to late December.

The Coorong, a wild place bordered by lagoon and ocean.

SOME BIRDS OF THE COORONG

Left and clockwise: A chorus line of Silver Gulls; migratory wading birds waiting for the tide to turn; Cape Barren Geese; Pied Oystercatchers.

FLINDERS CHASE NATIONAL PARK

Remarkable Rocks are eroded granite boulders, standing on a granite dome at Kirkpatrick Point.

ACCESS

By air from Adelaide. Car ferry from Port Adelaide to Kingscote or from Cape Jervis to Penneshaw. Bus tours and hire cars available. Unsealed roads within park.

CAMPING

Permit required and fees charged. Main facilities at Rocky River. Bush camping sites with toilets at Snake Lagoon, West Bay, and Harveys Return. Booking required for West Bay and cottages at Rocky River, Cape du Couedic and Cape Borda.

WHEN TO VISIT

Any time of year. Wildflowers from late July to November.

SAFETY

Swimming is not recommended.

Lichens colour the granite formations in vivid tones.

Admirals Arch frames an ocean view at Cape du Couedic.

Flinders Chase NP is the largest of Kangaroo Island's many conservation reserves, and is best-known for its abundant wildlife and striking coastal rock formations.

The park spreads across the western end of the island's gently undulating plateau which is cloaked in dense, windshorn mallee and heath scrub. Creeks and rivers fringed with eucalypt forest drop from the low plateau to small beaches tucked between eroded granite headlands and limestone cliffs.

Kangaroo Island wildlife owes its diversity to a lack of feral predators and the introduction of native animals thought to be under threat on the mainland. Koala, Platypus and ringtail possums now coexist with original residents, such as the Kangaroo Island Kangaroo, Tammar Wallaby and Short-beaked Echidna.

Wildlife encounters are common on the island. Kangaroos, possums and Cape Barren Geese show no fear of people and haunt picnic and camping areas in search of food, although feeding is discouraged by park authorities. Visitors can also take a guided walk through the Australian Sea-lion colony at nearby Seal Bay Conservation Park. Other frequently seen coastal inhabitants include seabirds, waders, Little Penguin and New Zealand Fur-seal.

A male Australian Sea-lion.

Watching Australian Sea-lions on the beach at Seal Bay Conservation Park.

Flinders Chase NP is well-serviced with gravel roads which lead to coastal lookouts and walking tracks. At Cape du Couedic, an historic lighthouse overlooks spectacular cliffs fronting the wild Southern Ocean. It is a short walk from the lighthouse to Admirals Arch, a natural limestone tunnel inhabited by fur-seals. A little to the east, the much-photographed Remarkable Rocks do a precarious balancing act on the granite dome of Kirkpatrick Point.

The 3 km Rocky River Trail at Snake Lagoon features mallee scrub and a delightful waterfall as it follows the river to a small sandy beach. Tracks alongside Sandy Creek and Breakneck River offer a look at the park's heath and forest habitats before emerging on secluded beaches.

The park's longest and most strenuous walk is the 8 km circuit at Ravine des Casoars, named after the island's now extinct dwarf emus. This 3 hour creekside walk through eucalypt forest in the north of the park includes large limestone caves where Little Penguins live.

SOME ISLAND WILDLIFE

Above and clockwise: Koala; Common Ringtail Possum; Little Penguin.

Driving through Flinders Chase NP.

Flinders Chase bushscape.

ACCESS

50 km from Hawker to Wilpena Pound on sealed road. Highway 83 provides access to the western side of the park between Hawker and Parachilna. Gravel roads in the park may be impassable after rain.

CAMPING

Privately-owned camping and caravan park at Wilpena Pound. Permit required and fees charged for basic facilities at northern camp sites. Conditions apply to bush camping.

WHEN TO VISIT

Popular from May to October, when temperatures are moderate. Nights can be cold.

SAFETY

Carry drinking water when walking or driving. Fireban season extends from November through April.

Wilpena Pound is the heart of the Flinders Ranges NP. This vast basin rimmed by jagged bluffs rises dramatically above grassy plains and cypress-clad foothills. Creeks lined with River Red Gums follow the steep-sided gorges which carve through the ranges.

The rich colours and striking beauty of the ranges are an inspiration to the park's many visitors. From dawn to dusk, sunlight paints the sheer cliffs and rocky outcrops in subtle colours. Spring rains bring a carpet of mauve, red and yellow flowers to the grasslands and lower slopes, and for many, this is time the ranges are at their most enchanting.

Rockpools and seasonal waterholes attract frogs, reptiles and colourful birds. Kangaroos graze on the plains, while Yellow-footed Rock Wallabies take refuge high in the rocky outcrops of the gorges.

Rock art sites, scattered throughout the park, are evidence of the area's importance to the Adnyamathanha people. Failed wheat and grazing ventures have left their mark among the ruins at Aroona and the Pound.

A network of gravel roads and walking tracks bring the park's natural and cultural features within easy reach of family groups, as well as providing access to some challenging treks for experienced bushwalkers.

Wilpena Pound, the massive crater that forms a centrepiece to the Flinders Ranges NP.

The Flinders Ranges rise from plains which, in summer, are covered with golden grasses.

Above: Unsealed roads take visitors through the spectacular landscapes of Flinders Ranges NP. *Below inset:* Sturt's desert pea.

Delightful creekside walks are a feature of Bunyeroo and Brachina Gorges. The Corridors Through Time trail in the limestone and quartzite gorge at Brachina is a self-guided tour through the park's geological history. In the north-east corner of the park, there is a 6 hour return walk at Wilka-willina where marine fossils, embedded in the gorge wall, indicate the region's sea bed origins.

Of the park's longer walks, the northern leg of the 1500 km Heysen Trail is the best known. The marked trail which begins on the Fleurieu Peninsula, south of Adelaide, makes a south–north traverse of the park before ending at Parachilna Gorge.

Short walks near the southern boundary lead to Aboriginal art sites at Sacred Canyon and Arkaroo where paintings depict the creation of Wilpena Pound. Tracks from the Pound carpark include a $1^{1}/_{2}$ hour historic walk to Hills Homestead. Add an extra hour to climb Wangara Hill behind the homestead for good views across Wilpena Pound. The 1 hour self-guided Drought Busters walk near Wilpena Creek explains how plants survive in the park's semi-arid environment.

The arduous ridgetop walk to the Flinders highest summit, St Mary Peak, is a 6–7 hour return trip through the Pound. A somewhat less strenuous trail, with equally spectacular views, is the 3 hour Mt Ohlssen Bagge summit walk.

The elusive Yellow-footed Rock-wallaby.

A magnificent river red gum.

Camping in the foothills of the ranges.

A grass-tree brandishes its spears aloft.

ACCESS

50 km W of Port Lincoln off the Flinders Hwy. 4WD N of Yangie Beach and Point Avoid. Boat ramps at Coffin Bay township and Point Avoid.

CAMPING

Fees charged at designated sites with limited or no facilities.

WHEN TO VISIT

Autumn and spring. Summer hot and dry; winters wet and windy.

SAFETY

Swimming and diving not recommended along seaward coastline.

On the eastern end of the Australian Bight, Coffin Bay is one of South Australia's top coastal getaways. The park covers a low peninsula of ocean-driven sand trapped between granite outcrops.

The sheltered northern shoreline weaves around Coffin Bay, creating channels, inlets and broad waterways ideal for boating, fishing, sailboarding and diving. Waders, heathland parrots and seabirds such as albatross, petrels and Osprey provide hours of birdwatching.

Across the saltpans, woodlands, shifting dunes and heath lie the wild, exposed south and west coasts. Heavy ocean swells, strong currents and prevailing westerlies sweep among rocky reefs and islands to shape the park's limestone cliffs, granite shelves and ocean beaches. Swimming and diving are not advisable on this side of the park, as coastal and island seal colonies attract sharks, including the White Pointer, which has a reputation for aggressive feeding, including on humans.

A gravel road provides conventional vehicle access to Gunyah Beach with its extensive dunefields and to the limestone headland at Point Avoid. Visitors can also get to bayside Yangie Beach with no difficulty. Beyond here it's 4WD and bushwalking territory.

A 4WD track winds through tea-tree woodland and coastal heath along the northern side of the peninsula taking in the Lake Damascus saltpan and Seven Mile Beach. The track splits after leaving the beach: one branch heads north to the popular fishing spot of Point Sir Isaac; the southern track ends at Sensation Beach. The wilderness area south of this dune-backed surf beach offers a range of full- and half-day walks.

A seascape in Coffin Bay NP.

The constant crashing of the Southern Ocean swells has shaped the limestone cliffs.

NORTHERN TERRITORY

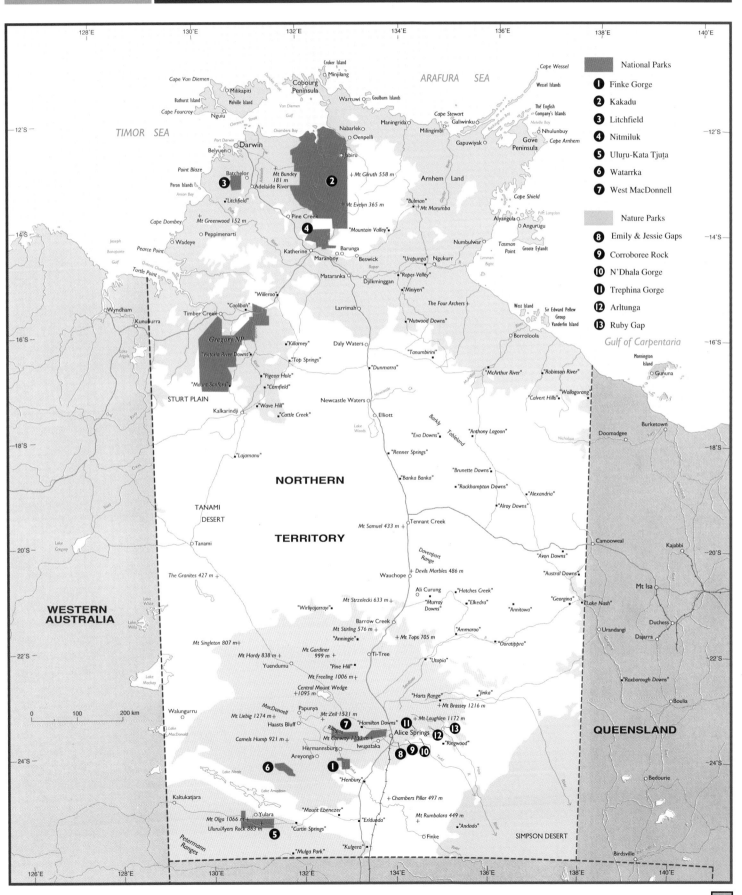

National Parks

1. Finke Gorge
2. Kakadu
3. Litchfield
4. Nitmiluk
5. Uluru-Kata Tjuṯa
6. Watarrka
7. West MacDonnell

Nature Parks

8. Emily & Jessie Gaps
9. Corroboree Rock
10. N'Dhala Gorge
11. Trephina Gorge
12. Arltunga
13. Ruby Gap

ULURU-KATA TJUTA NATIONAL PARK

ACCESS

450 km SW of Alice Springs. Conventional vehicle access via Stuart and Lasseter Hwys. Commercial flights and bus tours from Alice Springs. Sealed roads may close after heavy rain.

CAMPING

No camping in park. Range of camping and accommodation facilities at Yulara village, 5 km from park entrance.

WHEN TO VISIT

Generally stable weather from April through September with comfortable days and cold nights.

SAFETY

Carry drinking water when walking. Avoid the midday heat.

Colours are intense at Uluru.

Uluru is at its most colourful in the hours around dawn and sunset. Here it is framed by desert oaks. *Inset below:* Thorny Devil.

The monolithic rocks of Uluru and Kata Tjuta have come to symbolise Australia's desert heartland. These famous landmarks stand alone and defiant in the face of powerful elements that have reduced surrounding mountains to arid plains and sand dunes.

To the Anangu, the Aboriginal owners of the park, this remarkable landscape is a record of the journeys and activities of ancestral beings who shaped the land and created life. Anangu people are responsible for ensuring the well-being of the land and its lifeforms through an integrated system of spiritual beliefs and cultural laws known as Tjukurpa.

The significance of the park's cultural, biological and geological features is recognised internationally and it is both a World Heritage Area and a World Biosphere Reserve. Uluru, or Ayers Rock, 348 m tall and 9.4 km in circumference, is the tip of a great slab of sedimentary arkose thought to extend several kilometres underground. Many caves honeycomb its upper levels, gash its ribbed sides and undercut its base.

The Rock's near vertical slopes carry little vegetation, but dense thickets of bloodwood, wattle and fig cluster in gullies and sheltered sites around the base, especially near waterholes and creek beds. A sparse desert-tolerant cover of mulga, mallee eucalypts, desert oaks and tussock grasses stabilise the sandplains and dunes which stretch westward beyond Kata Tjuta.

The domes of Kata Tjuta are of great significance to the Aborigines of the region.

The park is well serviced with internal roads and day-use facilities. Information displays, viewing areas and walking tracks introduce visitors to desert ecosystems, ancient landforms and Aboriginal culture. The key to enjoying and appreciating this timeless land is to "take it slowly". Walkers are advised to carry water with them.

Signed lookouts along the main road between Yulara and Uluru provide good opportunities for sunset photography and visitors can get all-round views from the road encircling the Rock.

Park staff, Anangu guides and tour operators run a multitude of guided walks and activities. The Anangu-led Liru Walk is a very rewarding experience. Visitors can guide themselves around all or part of the 9 km Uluru Circuit Walk. An excellent booklet accompanies the Mala and Mutitjulu sections of the track offering an insight to the ages-old relationship between the land and the Anangu.

It is against Aboriginal law to climb Uluru, and the Anangu request visitors to refrain. However, visitors are not prevented from doing so. The summit views are certainly magnificent, but the 1.6 km climb is a strenuous 2 hour return trip and visitors should be aware of the potential risks. Health problems, and the expressed wishes of the owners, should make people think twice before attempting the climb.

The many-headed Kata Tjuta is a jumble of massive sandstone domes, narrow canyons and valleys. Mt Olga, 546 m above ground level, is the tallest of the 36 domes and also the highest point in the park.

Its varied topography supports a greater range of plants than Uluru, including river red gums and a delicate pink lily associated with damp sites.

Beyond the picnic and viewing area, two marked trails lead among the domes. The 2 km walk into Olga Gorge takes about 1 hour, while the 6 km circuit through the Valley of the Winds is a 3 hour venture.

Kata Tjuta is a good wildlife-watching area where birds of prey, wallaroo, and Black-footed Rock-wallaby are commonly seen. Among the many reptiles in the area are Perentie, Australia's largest lizard, and the ant-eating Thorny Devil.

A distant view of Kata-Tjuta, whose Aboriginal name can be translated as "many heads".

WATARRKA NATIONAL PARK

ACCESS

Conventional vehicle access on partly sealed roads. 310 km W of Alice Springs via Stuart Hwy and Ernest Giles Rd, or via Larapinta Dr. and Mereenie Loop Rd (passes through Aboriginal land; permit must be obtained in Alice Springs). 300 km N of Uluṟu via Lasseter Hwy, Luritja, Ernest Giles Rds. Check road conditions after rain.

CAMPING

Private camping ground and accommodation within the park. Permit required for bush camping on overnight walks.

WHEN TO VISIT

April through October. Summers are extremely hot.

SAFETY

Carry drinking water when walking. Permission required for off-track walks.

A sheer cliff in Watarrka NP.

Kings Canyon winds across a rugged landscape, its gorges carved up to 200 metres deep into the sandstone.

Watarrka NP lies on the western edge of the aridland central ranges midway between Alice Springs and Uluṟu. The layered sandstone bluffs of George Gill Range rise several hundred metres from the desert plains to a tabletop plateau gashed with sheer-walled gorges and narrow ravines.

Kings Canyon is the most spectacular of the park's many gorges which harbour deep waterholes and the descendants of long-gone tropical rainforests. Cycads, ferns and mosses grow among river red gum, cypress and grevillea in the moist, shady environment of the canyon's boulder-strewn floor.

Above the polished cliffs at the canyon head sits another gorge and a delightful oasis called the Garden of Eden. Figs and other rainforest species surround permanent rockpools in this beautiful hanging valley. Tree-frogs, brush-tailed possums and black-headed goannas shelter in these heavily wooded gorges. Following heavy rain, the rockpool overflows, cascading down the cliffs and into Kings Creek.

Clusters of bonsai ghost gums and sandstone domes resembling giant beehives create a surreal landscape on the plateau surrounding Watarrka's hidden valleys. This is the realm of ring-tailed dragons, wallaroos and rock wallabies.

Visitors can explore Kings Canyon on two well-marked walking tracks. The Kings Creek Walk is a 3 km return trip along the rocky creek bed to a lookout beneath the gorge's towering cliffs. The 6 km Kings Canyon Walk is a 4 hour meander around the gorge rim, featuring rock domes, stunning gorge views and a side track to the Garden of Eden.

Aboriginal art sites are the highlight of a 1.3 km walk to the Kathleen Springs rockpools. Such sites must be treated with respect. The track begins at a picnic area just beyond the visitor centre and is suitable for wheelchairs. Signs along the way explain the cultural significance of this small gorge.

Park staff can advise experienced bushwalkers about off-track routes through the George Gill Range, including the overnight trek between Kathleen Springs and Kings Canyon.

FINKE GORGE NATIONAL PARK

The Finke River has been carving its way through the sandstone ranges of the dry heart of central Australia for more than 65 million years.

During that time, glaciers and rainforests have come and gone. Mountain peaks have worn away to low hills. And the mighty Finke River which flowed through deep gorges to Lake Eyre now makes a dry run through a shallow meandering river valley before dissipating among the sand dunes of the Simpson Desert.

After heavy rain, the Finke displays some of its former gorge-carving power as raging floodwaters eat into its banks and cast debris high into the fringing river red gums. Periodic flooding replenishes the river system's permanent waterholes. In dry times, these important waterbird refuges are maintained by rain water filtering slowly through the porous sandstone ranges.

Palm Valley is the park's most accessible gorge, with lookouts and walking tracks revealing its hidden features. The intriguing valley opens to a natural amphitheatre of sculpted red cliffs near its junction with the Finke River. Upstream, the valley narrows to a palm-filled canyon etched with caves and lined with clear rockpools. Above the rock walls, the dry, stony ranges carry a sparse cover of spinifex, wattle and mulga in startling contrast to the cool, sheltered gorge.

Palm Valley contains unique plants such as this palm tree.

Bushwalkers can take a long trek from Palm Valley down the Finke Gorge through the Glen of Palms to Boggy Hole's permanent water.

Visitors seeking a 120 km, 5–6 hour 4WD adventure can traverse the park north–south in dry weather, following the beds of Ellery Creek and the Finke River, cutting through the James Range south of Boggy Hole, then continuing beyond the park boundary to Ernest Giles Road.

ACCESS

130 km W of Alice Springs via Hermannsburg. 4WD only for last 19 km along sandy river bed. Dry weather access.

CAMPING

Fees charged. Good facilities with showers at Palm Valley. Bush camping at Boggy Hole.

WHEN TO VISIT

April through October. Summer is uncomfortably hot.

SAFETY

Carry drinking water when walking. 4WD route through the park requires experience and extreme care. Ensure adequate fuel. Check track conditions with ranger.

The Finke River creates oases such as the one shown here in the arid country of Finke Gorge NP.

Spinifex Pigeon.

WEST MACDONNELL NATIONAL PARK

ACCESS

The park extends 170 km W from the outskirts of Alice Springs. Sealed access via Larapinta and Namatjira Drives. 17 km sealed bicycle path from Flynn's Grave on Larapinta Drive to Simpsons Gap. Simpsons Gap open from 8 a.m. to 8 p.m.

CAMPING

Fees charged. Basic facilities at Ellery Creek Big Hole, Serpentine Chalet, Ormiston Gorge and Redbank Gorge. Privately-owned facilities at Glen Helen Gorge include caravan sites. Permit required for bush camping on the Larapinta Trail.

WHEN TO VISIT

April through October is most comfortable for walking.

SAFETY

Carry drinking water when walking.

A tawny-yellow Dingo.

A ghost gum rises in front of the West MacDonnell Ranges.

The rugged features of the West MacDonnell Ranges dominate the Northern Territory's central range country. Extending for more than 200 km west of Alice Springs, great folds of sedimentary rock stand above arid spinifex plains in long curving walls separated by narrow valleys. Ancestral tributaries of the Todd and Finke Rivers have cut through the multi-coloured layers of quartzite, clay, siltstone and coal, leaving steep-walled gaps in the ranges.

Rivers no longer flow through the West MacDonnells except during floods, but deep permanent waterholes lie along the sandy creek beds. Many are found at the base of precipitous rock walls which shelter fig trees, cycads and delicate ferns. The pools teem with aquatic animals and attract many waterbirds and large mammals, such as rock-wallabies and wallaroos.

These oases of permanent water and abundant food have been a meeting place for Aboriginal people for thousands of years. The waterholes, canyons and surrounding landforms are also the source of many central Australian Aboriginal creation beliefs.

Guided walks, interpretive signs and walking trails provide a rewarding introduction to the park's more accessible gorges and cultural sites.

Just 23 km from Alice Springs, the impressive red cliffs of Simpsons Gap overlook Roe Creek's broad sandy watercourse and a small waterhole where rock-wallabies gather at dawn and dusk.

Ormiston Gorge and Pound was created by massive upheavals within the Earth. Its walls were then eroded by wind and water.

At Standley Chasm, vertical rock walls almost meet overhead allowing only the midday sun to reach the canyon floor. Visitors can take a 1.5 km creek-bed walk through the chasm.

Further west, at Serpentine Gorge, there is a half-hour walk to a deep waterhole at the entrance to this narrow, twisting gorge. A clifftop lookout offers excellent views of the gorge and surrounding ranges.

Ormiston Gorge, about 130 km west of Alice Springs is considered the park's most spectacular natural feature. River red gums shade a series of waterholes beneath imposing rock walls which open to a 10 km wide, cliff-rimmed amphitheatre.

The main waterhole is a popular swimming spot and only a 15 minute walk from the carpark. Walking trails include the 3 hour Pound Walk that follows the cliff edges for panoramic views over Ormiston Pound, returning to the main rockpool through the basin.

Simpsons Gap, one of the more popular MacDonnell Range gorges.

The ranges are of significance to Aboriginal people.

For many thousands of years, Aborigines used pigments from the Ochre Pits for ceremonies and trade.

A visit to the West MacDonnells should include a side trip to the Ochre Pits. The red, white and yellow soils that stripe the ravine walls provided a rich palette of ochres for Aboriginal ceremonial use and a highly valued trading commodity.

Visitors can also enjoy refreshing waterholes and gorge scenery at Ellery Creek Big Hole, Glen Helen Gorge and the more remote Redbank Gorge.

Eventually the park's major attractions will be linked by the Larapinta Trail, a 220 km walking track along the backbone of the West MacDonnell Ranges between the Alice Springs Telegraph Station and Mt Razorback. Completed sections offer overnight walks ranging from easy grade family walks to difficult hikes through isolated regions.

Glen Helen Gorge, where pools of water ensure plentiful animal and plant life.

ACCESS

East of Alice Springs via Ross Hwy and unsealed Arltunga Tourist Drive (may be impassable after heavy rain). It is 15 km to Emily and Jessie Gaps Nature Park; 41 km to Corroboree Rock Conservation Reserve; 71 km to Trephina Gorge Nature Park; 79 km to N'Dhala Gorge Nature Park (4WD only); 107 km to Arltunga Historical Reserve and 145 km to Ruby Gap Nature Park (4WD only).

CAMPING

Basic facilities at Trephina Gorge and N'Dhala Gorge. Bush camping with no facilities at Ruby Gap.

WHEN TO VISIT

April through October. Summer is extremely hot. Avoid unsealed roads after heavy falls of rain.

SAFETY

Carry drinking water. Avoid walking in the heat of the day.

Sturt's desert rose.

The physical beauty and intriguing history of the MacDonnell Ranges extends eastwards beyond Alice Springs. It is an arid, rock and spinifex landscape where tributaries of the Todd and Hale Rivers snake through the East MacDonnell Ranges along dry creek beds. River red gums follow their sandy progress to permanent waterholes in river-cut gorges.

The ranges retain great cultural significance for the Eastern Arrernte Aborigines, having been a source of food, water and spiritual beliefs for countless generations. Homesteads and mines are evidence of the area's historical role in the European search for pastoral and mineral wealth during the late 1800s – some successful, some abandoned.

Several small parks and reserves along the Ross Highway and Arltunga Tourist Drive highlight many natural and cultural features in the East MacDonnell Ranges.

Emily and Jessie Gaps Nature Park, 10 km from the Stuart Highway, consists of 695 ha and offers a range-top walk between the two gaps where semi-permanent pools lie in sandy creek beds. Emily Gap is a sacred site with a large rock painting of the Caterpillar Dreaming. Euros and rock-wallabies can be seen drinking at dawn and dusk.

Corroboree Rock Conservation Reserve is 7 ha in extent. A self-guided walk can be made around a sandstone outcrop of Aboriginal significance.

Trephina Gorge Nature Park contains the most spectacular gorge in the East MacDonnells. Visitors can take the 4WD track or walk across hill country to John Hayes Rockhole, or explore five walks which take from 45 minutes to 6 hours.

The MacDonnell Ranges stretch for 400 km.

Trephina Gorge, which is 75 km from Alice Springs, can be explored on a series of walking tracks.

N'Dhala Gorge Nature Park covers only 501 ha but offers thousands of Aboriginal rock engravings. It has two scenic gorges and features stands of native pine on exposed hill slopes.

Arltunga Historical Reserve offers relics of the discovery of gold in 1887, including a ghost town in which a restored police station and lock-up and a mine are open to the public.

The discovery of "rubies" which turned out to be garnets gave Ruby Gap Nature Park its name in 1885. It is one of a number of gorges along the Hale River, and consists of a winding corridor of cliffs leading to Glen Annie Gorge waterhole. At 9257 ha in extent, it is one of the East MacDonnell's largest reserves.

Before setting off to explore either the East or the West MacDonnell Ranges, it is worth visiting the Olive Pink Flora Reserve and the Desert Park at Alice Springs. Here an insight can be gained into the plants and animals which live in some of Australia's most testing environments.

Emily Gap is in an area with special significance to Aboriginal people.

ARIDLAND ANIMALS

A rocky escarpment in the East MacDonnells.

Desert animals often shelter during daytime heat. The Black-footed Rock-wallaby (left) dozes beneath a rocky overhang; the Bilby (above) retreats to a burrow; and the Marbled Gecko (below) hides in a rock crevice.

Ghost gums tap underground water.

ACCESS

Sealed access roads off the Stuart Hwy. The gorge is 30 km NE of Katherine and the Edith Falls turnoff is 42 km from the town.

CAMPING

Privately-owned camping grounds at Katherine Gorge and Edith Falls. Permit required for bush camping at designated sites on the plateau and along the river.

WHEN TO VISIT

May through September.

SAFETY

Bushwalkers planning overnight hikes must register at the visitor centre. Freshwater Crocodiles are harmless unless provoked.

Above: Nitmiluk NP centres on deep gorges carved into the Arnhem Land Plateau by the Katherine River. *Inset below:* Water-lily.

The Katherine River snakes through the middle of Nitmiluk NP in a series of long, deep reaches beneath the high cliffs of the magnificent sandstone Katherine Gorge.

Boulder rapids separate the 13 gorges which, in the Dry, rise from tranquil stretches of the river. Freshwater Crocodiles bask on rock ledges and sand banks. Breaches in the gorge walls reveal narrow canyons headed with waterfalls and filled with palms, figs and paperbarks.

The Arnhem Land Plateau stretches away on either side of the gorge, extending north to Kakadu. Its rocky escarpments are the haunt of wallaroos, wallabies, lizards and snakes. This harsh, dry country carries spinifex grass, sparse eucalypts and hardy shrubs which attract colourful parrots during the winter flowering season. In summer, monsoon rains wash over the plateau, swelling the river into flood.

Cruise boats, canoes and other small craft take sightseers on tours of Nitmiluk NP's gorges.

During the dry season, the river subsides into a series of pools separated by rocky bars.

Honeyeaters feed on blossoms along the watercourses.

The Nankeen Night Heron fishes after dark.

North of the gorge, the western face of the plateau is cleft by another canyon where the Edith River drops through a succession of waterfalls and rockpools. The large pool below Edith Falls, with its palm and monsoon rainforest backdrop, is very popular with day visitors, but there are several delightful rockpools only a short walk beyond the falls.

There are more than 100 km of marked trails in the park, ranging from short walks along the Katherine and Edith Rivers to a 66 km wilderness trail across the plateau between the two.

Linked tracks at Katherine Gorge offer a variety of full-day and overnight hikes which wind over the plateau dropping down to the gorge through side canyons. However, walking on the rugged plateau is arduous and most visitors prefer to see the gorge by boat.

Tour boats operate daily in the five lower sections of the gorge, while canoeists can paddle their way upstream for more than 30 km, carrying their canoes over the rock bars that separate the reaches. For a bit of fun and excitement catch a lift upstream on one of the tour boats then float back through the gorge on an air mattress.

Eucalypts and native grasses feature in the landscapes above the gorges.

LITCHFIELD NATIONAL PARK

ACCESS

About 100 km S of Darwin. Conventional vehicle access on partly sealed roads via Stuart Hwy and Batchelor or Cox Peninsula and Litchfield Park Rds. Towing a caravan or trailer is not recommended. 4WD tracks may be closed after heavy rain. Commercial tours from Darwin.

CAMPING

Fees apply. Basic facilities at Wangi Falls, Florence Falls and Buley Rockhole. 4WD access only to Tjaynera Falls camping area. Permit required for bush camping on overnight walks.

WHEN TO VISIT

Comfortable walking conditions from May to October. Spectacular waterfalls during the wet season from November through April.

SAFETY

Carry drinking water.

Pied Imperial-Pigeon

Litchfield NP encompasses more than 146 118 ha of sandstone highlands, raw-faced escarpments and lowland floodplains. Each wet season, tributaries of the Adelaide, Finniss and Reynolds Rivers rumble over the edges of Tabletop Range, swelling rockpools and creek beds with monsoon rains before spilling across paperbark floodplains.

The park's many waterfalls still put on an impressive display during the dry season when cool, deep rockpools offer welcome relief from the Top End heat.

Monsoon rainforest in a sheltered gorge in Litchfield NP.

Wangi Falls, where cool water tumbles over the sandstone of the Tabletop Range into a pool surrounded by forest.

Litchfield is particularly blessed with birdlife. The jabiru (or Black-necked Stork) is among the waterbirds that inhabit the floodplains during the Wet. The Red-tailed Black-Cockatoo and Sulphur-crested Cockatoo are common sights.

Creekside pockets of monsoon rainforest can be found along the base of the western escarpment. Palms, ferns and looping vines grow beneath semi-deciduous trees which shed their broad leaves when water-stressed. In summer, colourful fruit supplements the diet of rainforest animals, such as the Northern Brushtail Possum, Black Flying-fox and Pied Imperial-Pigeon.

On the plateau above the escarpment, open grasslands with occasional boggy patches grow between eucalypt woodlands supporting an understorey of sand palms, cycads, banksias, wattles and grevilleas.

Tolmer Falls, a picturesque feature of Litchfield NP.

Buley Rockhole is a popular cool spot in Litchfield NP.

The Tabletop is good birdwatching country, especially in summer when woodland flowers attract noisy flocks of Red-winged Parrots, Northern Rosellas and Rainbow Lorikeets. The plateau also features a "lost city" of weathered sandstone pillars and giant termite mounds aligned within a few degrees of true north–south. Colonies of nature's ultimate recycler, the termite, live in vast tunnel systems under these apparent climate control towers.

The Litchfield Park Road cuts across the park providing access to the plateau and several waterfalls along the western escarpment. Florence, Wangi and Tolmer Falls, as well as the Buley Rockhole cascades, are popular swimming spots throughout the year. Lookouts and marked walking tracks offer great waterfall views.

Visitors will need a 4WD to reach the Lost City and Tjaynera Falls where Sandy Creek drops over sandstone cliffs into a valley dominated by tall paperbarks. There is an excellent patch of monsoon rainforest not far downstream from the falls.

A rough 4WD track continues south along the edge of the western floodplains, crossing Reynolds River and the park's southern boundary before joining Daly River Road. Another 4WD route winds through the Adelaide River catchment in the eastern region between Milton Road and Daly River Road. Both routes are subject to flooding during the wet season.

These termite mounds are aligned so they get the least possible exposure to hot summer sunlight.

Florence Falls is a popular place to picnic and swim.

KAKADU NATIONAL PARK

ACCESS

145 km from Darwin via the Arnhem Hwy or 59 km from Pine Creek on the Kakadu Hwy. Wet season road closures. Tours operate from Darwin.

CAMPING

Fees charged. Seasonal closures. Good facilities at Mardugal, Muirella Park and Gunlom. Limited facilities at other designated sites. Permit required for camping on overnight bushwalks. Range of accommodation within the park.

WHEN TO VISIT

June through August is most comfortable. Wet season from late December to April.

SAFETY

Take care near water — Saltwater Crocodiles are dangerous. Carry drinking water.

Frilled Lizard, common in Kakadu.

Ubirr Lookout, in the far north-east of Kakadu NP, is a splendid place from which to view the floodplain of the East Alligator River.

Kakadu lies at the heart of Northern Territory's monsoon country. This World Heritage listed park covers nearly 2 million hectares of rugged sandstone country and river floodplains. From late December through March, heavy rains wash across the Arnhem Land Plateau, over-filling the Alligator River systems and flooding the low-lying coastal plains.

As monsoon rains replenish the park's wetlands, countless waterbirds begin their mating rituals. By winter, raging waterfalls have become silver arcs dropping to permanent rockpools, and the waterbirds retreat from the drying plains to crowd the edges of shallow billabongs and swamps.

Birds such as this jabiru flock to the floodplain in the Wet.

Kakadu's international reputation extends to from its wetlands to its galleries of Aboriginal rock art. A remarkable record of human life spanning more than 40 000 years has been etched and painted on the caves, overhangs and outliers of the plateau's 500 km long western escarpment. Although the descendants of the original artists no longer renew the paintings, the lives of the Bininj people, who share the management of the park, still revolve around the spiritual and cultural laws associated with these ceremonial sites.

The mighty sandstone Arnhem Land Escarpment.

Life for the park's Aboriginal owners is also dictated by the changing seasons. Each indicates the readiness of plant and animal food sources in habitats ranging from the tidal mudflats of Van Diemen Gulf to the hardy shrubs and grasses on the plateau. In between lie mangrove forests, sedgeland plains, paperbark swamps, eucalypt woodlands and monsoon rainforests.

Kakadu's wetlands and sandstone escarpments are most spectacular during the wet season. However, access is usually limited because of flooding which causes road closures. Access may be restricted, too, to limit damage to the environment. Most people visit from May through August when temperatures are cooler and before the floodplains dry out.

Jabiru, the main town in the park, is about 200 km along the Arnhem Highway from the Stuart Highway turnoff. Sealed roads take visitors to many of the park's popular features, but some roads are suitable for 4WD vehicles only.

The massive bulk of Nourlangie Rock, seen over Anbangbang Billabong.

Jim Jim Falls becomes a thundering torrent in the Wet.

Kakadu flood plains after the Wet breaks.

Birdwatching on Yellow Water.

There is an excellent visitor centre at the park headquarters at Bowali on the Kakadu Highway near Jabiru. Maps of the park and information on road conditions, walking trails, flora and fauna, and tours can be obtained. During the dry season, park rangers conduct guided walks. Commercial operators run scenic flights, 4WD and boat tours within the park.

At Yellow Water is the Warradjan Aboriginal Cultural Centre, featuring an outstanding display of Bininj cultural materials and information to help visitors better understand their way of life.

Kakadu can be roughly divided into five districts, each with centralised facilities and walking tracks located near the Arnhem or Kakadu Highways. These include South Alligator, East Alligator, Nourlangie, Jim Jim and Mary River.

Birdwatching in a paperbark swamp.

Some rock art in Kakadu was continually renewed for many thousands of years.

Each district is unique, whether it be the billabongs, rivers and wetlands of the floodplains or the rock art, waterfalls and forests of the escarpment.

The plateau is harsh, isolated country, only to be tackled by experienced bushwalkers. A special permit and advice on exploring this area can be obtained from park headquarters. Walking in the middle of the day should be avoided. Early mornings and late afternoons are the best times for people (and wildlife) to be about.

Three 4WD tracks between 60 and 80 km long take well-prepared visitors beyond the main tourist areas beside the floodplains to the coast, along the floodplains between the Arnhem Highway and the Old Jim Jim Road, and to Jim Jim Falls and Twin Falls gorge.

Twin Falls is seen at its magnificent best during the summer Wet.

WATCH OUT FOR CROCODILES

The Saltwater Crocodile (left) lives in the sea and in tidal rivers. It can grow to 6 m in length, eats anything it can subdue and is potentially dangerous to humans. It is a good reason to avoid swimming in salt or brackish water in northern Australia. The Freshwater Crocodile (right) can grow to 3 m, lives in freshwater creeks and billabongs, eats fish and, unless harassed, is not dangerous to humans.

WESTERN AUSTRALIA

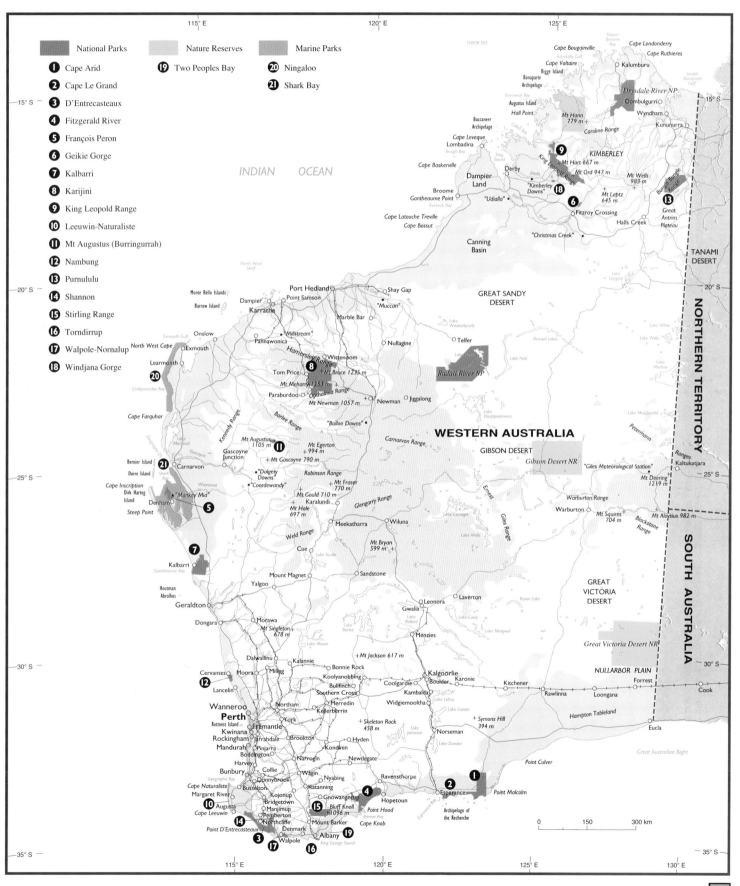

National Parks   **Nature Reserves**   **Marine Parks**

- ❶ Cape Arid
- ❷ Cape Le Grand
- ❸ D'Entrecasteaux
- ❹ Fitzgerald River
- ❺ François Peron
- ❻ Geikie Gorge
- ❼ Kalbarri
- ❽ Karijini
- ❾ King Leopold Range
- ❿ Leeuwin-Naturaliste
- ⓫ Mt Augustus (Burringurrah)
- ⓬ Nambung
- ⓭ Purnululu
- ⓮ Shannon
- ⓯ Stirling Range
- ⓰ Torndirrup
- ⓱ Walpole-Nornalup
- ⓲ Windjana Gorge
- ⓳ Two Peoples Bay
- ⓴ Ningaloo
- ㉑ Shark Bay

ACCESS

260 km S of Perth. Numerous access points between Dunsborough and Augusta.

CAMPING

Fees charged. Basic facilities at Conto, Point Rd and Boranup Forest.

WHEN TO VISIT

Summer for beach activities. August through October for wildflowers and whale watching.

SAFETY

Westerly winds generate huge swells in winter. Take care when walking near cliff edges.

Sugarloaf Rock is a haven for seabirds, including the Red-tailed Tropicbird, which breeds there in summer.

The lighthouse on Cape Leeuwin.

A prominent granite ridge, capped with limestone and sand dunes, forms the backbone of Leeuwin-Naturaliste NP. Wind, rain and ocean waves have eroded its western edges into an imposing coastline of rugged cliffs, rocky outcrops and platforms, secluded bays and long beaches.

Leaching rainwater has riddled the area with caves, tunnels and sinkholes. The fossilised remains of extinct animals litter the floors of some of these intriguing cavities, while subterranean streams run through others. A number of the larger caves have spectacular formations and are open for public viewing.

Salt-tolerant heath, growing on the seaward rises, comes into flower during spring, providing feasts of nectar for honeyeaters and small mammals. Shrubby woodlands run down the sheltered eastern slopes to tall forests of jarrah, karri, tuart and marri.

Heathland shrubs flower profusely in spring.

Spider orchids are bushland gems.

Limestone formations at Cape Leeuwin.

Western Grey Kangaroos, quolls, bandicoots, Honey-possums and ringtail possums can be seen, but the most conspicuous creatures are waders and seabirds. Each summer, flocks of migratory species join local bird populations: Sugarloaf Rock is famous among birdwatchers for its nesting Red-tailed Tropicbirds.

There's plenty of access to the park's 120 km coastline, which offers great surfing, fishing, diving and whale watching when Humpbacks pass by on their annual migration. A dive trail at Hamelin Bay takes an underwater look at four of the area's numerous shipwrecks.

A range of coastal walking trails feature scenic lookouts, historic sites and wildflower heathlands. Enthusiastic bushwalkers can do an extended ridge walk between the two capes. Inland attractions include karri forest walks, Boranup Lookout for coastal and hinterland views, and many fascinating limestone caves. Yallingup, Mammoth, Jewel and Lake Caves are open to the public, and experienced cavers can obtain a permit to explore others.

Regrowth karri forest will, in time, tower like the giants of old.

Observing seabirds from a lookout.

D'ENTRECASTEAUX NATIONAL PARK

ACCESS

380 km S of Perth via Pemberton and Northcliffe. 4WD tracks within park.

CAMPING

Camping permitted at Lake Jasper, Donnelly River and Windy Harbour. Conditions apply.

WHEN TO VISIT

Spring and summer.

SAFETY

Swells and rips make ocean treacherous.

D'Entrecasteaux NP is a large, undeveloped area that stretches along the coast east from Cape Beaufort to join Walpole-Nornalup NP at Mandalay Beach. Sandy beaches, interrupted by rugged headlands, sweep in front of extensive coastal dunes. Beyond lie sedgeland plains, freshwater lakes, swamps and karri-forested granite hills.

Canoeing is a good way to explore Lake Jasper and also the Donnelly River. Further east, at Yeagarup Lake, visitors can walk up onto huge inland sand hills. Bare of vegetation, the unstable sand is creeping slowly across the plains.

It's worth stopping at Mt Chudalup on the drive to Windy Harbour. A 1 km walking trail climbs to the top of this massive granite outcrop for 360° views of forests and heathlands that burst into bloom in spring. Visitors can drive from Windy Harbour to the clifftop lookout at Point D'Entrecasteaux or skirt behind the headland for a swim at Salmon Beach.

The park's wild coastline is largely the realm of 4WD adventurers. Several rough, often boggy, tracks cross the heathlands to long beaches and river estuaries well-known for good fishing.

Rugged coastal scenery in D'Entrecasteaux NP.

SHANNON NATIONAL PARK

ACCESS

358 km S of Perth via Manjimup on the South Western Hwy.

CAMPING

Fees charged. Good facilities with powered sites at the Old Shannon Town site.

WHEN TO VISIT

Any time of year. January and Easter are busy periods.

SAFETY

Expect cold weather and rain in winter.

This 53 500 ha park encompasses most of the Shannon River catchment. The Shannon begins its seaward journey in the heart of the south-west karri country. Emerging from the granite hills, the river crosses sedgeland flats and heath-covered dune systems to Broke Inlet, where it empties into the Southern Ocean between jagged limestone cliffs.

Camping and day-use facilities are centred around the abandoned mill town of Shannon. Information displays take an historical look at the town's role in the timber industry. The nearby dam is popular for swimming, canoeing and seasonal fishing. Visitors can walk through part of the river valley on the 3.5 km Shannon Dam trail or take in the views from the 5.5 km Rocks Walktrail as it climbs karri-clad granite slopes. A one-way scenic drive which loops around the Shannon River headwaters offers an in-depth look at old and regrowth karri forests.

Visitors can also drive down the western side of the park to coastal heath and wetland habitats at Broke Inlet. This huge shallow inlet is great for birdwatching, with mobs of seabirds and waders.

A road through a stand of karri in Shannon NP.

Walpole-Nornalup NP contains impressive karri and tingle forests. These giant hardwoods cloak the park's granite hills and sheltered valleys and follow the Deep, Frankland and Bow Rivers down to the edges of broad, shallow inlets.

In damp gullies, mosses and ferns flourish beneath the forest canopy, giving the impression of an eastern temperate rainforest. Where the tall canopy is more open, wattles, pea bush, hovea and creepers create an impenetrable understorey.

At the Valley of the Giants, a treetop suspension walkway and forest floor boardwalk introduce visitors to a rare stand of red tingle. Both walks are suitable for wheelchairs. At Hilltop and the Knoll, scenic drives and short walking tracks take visitors through remnant forests to lookouts over the Walpole and Nornalup Inlets.

Amongst the red tingle trees in the Valley of the Giants.

Early morning reflections in Nornalup Inlet.

This tree-top walk allows a view of the karri forest canopy.

Walkways give access to beaches through the dunes.

In the west, the forested hills slide into the undulating coastal heaths and peat swamps of the Nuyts Wilderness area. This undeveloped region is accessible only on foot, from a bridge across Deep River, just off Shedley Drive.

Vehicle access to the park's coastline is restricted to the eastern end between Conspicuous Cliff and Peaceful Bay. However, boats can be launched from Nornalup Inlet, and keen walkers can trek across the wilderness area to isolated bays around Point Nuyts.

Canoeing is a great way to explore the inlets and lower river reaches. During spring, when water levels are high, canoeists can travel the Frankland River to Circular Pool, where the river drops into a large rockpool.

Bushwalkers looking for an extended hike can follow a section of the 840 km Bibbulmun Track along Deep River, then continue west of the inlets to Walpole, and on through the Valley of the Giants.

ACCESS

430 km SE of Perth on the South Western Hwy.

CAMPING

Fees charged for sites at Crystal Springs. Conditions apply to bush camping. Privately-owned camping grounds at Rest Point, Coalmine Beach and Peaceful Bay.

WHEN TO VISIT

Winter for whale watching. Spring for wildflowers and canoeing. Summer for beach activities.

SAFETY

Watch for changing weather and tides on the coast.

Cowslip orchids.

ACCESS

10 km S of Albany on Frenchman Bay Rd.

CAMPING

No camping in park. Camping and caravan grounds at Albany and Frenchman Bay.

WHEN TO VISIT

Spring through autumn.

SAFETY

This is a dangerous coastline with big ocean swells.

The Salmon Holes, one of the loveliest places in Torndirrup NP, is popular with surfers.

Wind and ocean waves have battered the Torndirrup coastline into a scenic extravaganza of perpendicular cliffs, narrow canyons, stacked headlands and dune-backed bays.

The park covers 3880 ha of a double-pronged peninsula. One protective arm sweeps north around Princess Royal Harbour, and the other extends eastward between King George Sound and the Southern Ocean. The granite-derived sandy soils carry open eucalypt woodland, stunted coastal heath and pockets of karri, providing food and shelter for pygmy-possums, honeyeaters, kangaroos and a variety of snakes.

Most of the park's well-known features lie along its exposed southern face. Lookouts, over the Gap, Natural Bridge and the Blowholes, are within a few hundred metres of the carpark. Stony Hill, the highest point in the park, offers northern views, over the harbour and sound, to the Porongurup and Stirling Ranges. There are more good views from the Salmon Holes lookout across to Bald Head, and a set of steps down to a beach popular with surfers. The Bald Head trail is a strenuous 10 km return walk, via Limestone Head, which takes about 6 hours.

The ocean gnaws at the coast.

The Gap, where relentless waves have hollowed an opening beneath massive granite blocks.

STIRLING RANGE NATIONAL PARK

The Stirling Range is a 65 km chain of high, jutting peaks set amidst the south-west's cultivated plains. Hardened ocean sediments have folded, uplifted, cracked and eroded into a rugged mountainous landscape quite unlike the rest of Western Australia.

Dense prickly heath climbs the gullied foothills to a backbone of sharp-edged ridges, sheer bluffs and ragged summits, some exceeding 1000 m. Between the peaks lie old river valleys with stands of jarrah, wandoo, marri and other eucalypt forest trees.

Rain-laden sea winds create a cool, moist environment on the range, supporting over 1000 varieties of plants. Many, such as the Mondurup bell and Stirling Range mallee, are restricted to the park. Late winter through spring is the best time to see the park's famous wildflowers, including 123 different kinds of ground orchid.

Well-maintained roads offer great scenic drives through the park, and access to plenty of wildflower nature walks and summit trails. One of the most popular walking tracks climbs to the 1073 m summit of Bluff Knoll. This 6 km return walk takes 2–3 hours, with lots of stops to admire the wonderful views and flowering plants.

Mt Trio, Mondurup, Toll and Talyuberlup Peaks are other moderate 2–3 hour climbs. Toolbrunup Peak, with magnificent 360° views from its 1052 m summit, is a more difficult walk needing 3–4 hours to complete.

Stirling Range NP also offers some of the State's best off-track adventures. They are not for novices, but for experienced bushwalkers and rock climbers. There are well-known but unmarked routes in the eastern section, including a challenging 3 day ridge walk between Bluff Knoll and Ellen Peak.

The Stirling Range is noted for spring wildflowers.

ACCESS

About 80 km N of Albany or from Cranbrook via Salt River Rd.

CAMPING

Fees charged for facilities at Moingup Springs. Bush camping permitted on extended bushwalks. Commercial camping and caravan ground near northern entrance on Chester Pass Rd.

WHEN TO VISIT

Spring and autumn for bushwalking. September–December for wildflowers.

SAFETY

When walking, be prepared for rain and sudden drops in temperature.

TWO PEOPLES BAY NATURE RESERVE

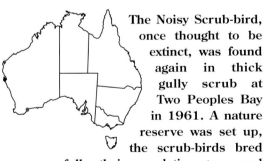

The Noisy Scrub-bird, once thought to be extinct, was found again in thick gully scrub at Two Peoples Bay in 1961. A nature reserve was set up, the scrub-birds bred successfully, their population grew, and breeding pairs have been relocated in similar habitats along the coast.

The reserve extends along a hilly promontory separating the bay from King George Sound. Reedy swamps lie behind the beach dunes, and dense scrub fills the drainage lines between sandy, heath-covered ridges. The survival of the ground-dwelling Noisy Scrub-bird determines which areas are open to the public, however, a 2 km nature circuit gives some idea of the habitats being protected in this 4640 ha nature reserve. There is always a chance of spotting, or

hearing, the scrub-bird or the equally rare Western Bristlebird.

There is easy access to the beach, where visitors can enjoy safe swimming or explore the coastline by boat during calm weather.

Enjoying the peace of Two Peoples Bay Nature Reserve.

ACCESS

20 km E of Albany.

CAMPING

None. Boat launching facilities.

WHEN TO VISIT

All year.

SAFETY

Do not enter dense vegetation.

ACCESS

180 km NE of Albany or 230 km W of Esperance. Conventional vehicle access on unsealed roads from Bremer Bay and Hopetoun. Designated 4WD tracks within park.

CAMPING

Fees charged for camping at designated sites, including Four Mile Beach. Permit required for bush camping on overnight walks.

WHEN TO VISIT

Late August to December for wildflowers and whale-watching.

SAFETY

4WD tracks can become boggy. Take drinking water.

Sea, surf and rocks.

Fitzgerald River NP is a large, undeveloped paradise on the south coast of Western Australia. The steep, rocky slopes of the Barren Ranges back a 100 km arc of magnificent coastline, featuring low sea cliffs, jagged headlands, pristine beaches and sheltered inlets.

Several rivers, including the Fitzgerald and Hamersley, have cut deep valleys between the ranges. Swamp yate and mallee woodlands mark their routes across extensive sandplains to the Southern Ocean. At Fitzgerald Inlet, intriguing red- and orange-banded cliffs of spongolite flank the river as it meanders into a placid estuarine lagoon favoured by seabirds and waders.

This World Biosphere Reserve is renowned for its diversity of plant life, especially in the sandy heathlands. Over 1700 species have been recorded, with about 15% found nowhere else. Throngs of honeyeaters and parrots revel among spring wildflowers such as royal hakea, Qualup bell and scarlet banksia *(inset)*. Two excellent scenic drives provide access to coastal, heathland and summit walking trails in the south-western and eastern sections of the park.

The hilltop climbs are not difficult, with peaks averaging 300 m above sea level, and the coastal views are superb. The western ridge climb at East Mt Barren is a 3 hour return hike, while the ascent of cone-shaped West Mt Barren takes about 2 hours return.

The rugged seashore is the icing on the park's cake. There, endless enjoyable hours of fishing, beachcombing and birdwatching, are topped off with whale-spotting in late winter and spring.

There are many enjoyable coastal walks, ranging from the 1 hour Point Ann Heritage Trail to a testing 22 km trek from the point to Twin Bays via Fitzgerald Inlet. 4WD access to Whale Bone Beach and Quoin Head extends the possibilities for walks between the many bays.

An unspoiled beach and a rocky headland in Fitzgerald River NP.

Cape Le Grand NP, within easy reach of Esperance, is a coastal retreat that is popular with beach and fishing enthusiasts. There are sheltered bays, sweeping beaches and granite headlands creating some impressive coastal scenery.

In the south-west corner, a chain of rounded granite hills peak at 350 m before sliding into the ocean at Cape Le Grand. Heath-covered sandplains roll across most of the park, interspersed with paperbark-fringed swamps. Wildflowers appear throughout the year, but September to November is the peak season. There are a number of short walks through sandplain heaths and hillside woodlands with great displays of bottlebrush, tea tree and spectacular banksias.

Visitors can get a good overview of the park and the Recherche Archipelago from the top of Frenchman Peak. This easy grade walk takes about 2 hours and features a huge wave-cut cavern below the 262 m summit.

The park's best-known walk is the 15 km Coastal Trail between Le Grand Beach and Rossiter Bay. Five access points divide the trail into easily managed sections of varying difficulty. Highlights include good birdwatching on the Rossiter Bay stretch, a self-guided heritage trail between Lucky Bay and Thistle Cove, and some challenging rock scrambling on the way to Hellfire Bay.

Boats can be launched from Lucky Bay providing access to isolated coves and nearby islands, but coastal waters can be treacherous and there are landing restrictions on some of the archipelago islands.

Rocky coast in Cape Le Grand NP.

ACCESS

Conventional vehicle access 40 km E of Esperance via Merivale and Cape Le Grand Rds. Boat ramp at Lucky Bay.

CAMPING

Fees charged. Basic facilities at Le Grand Beach and Lucky Bay. Permit required for bush camping.

WHEN TO VISIT

All year. Wildflowers best September–December.

SAFETY

Coastal waters can be treacherous.

CAPE ARID NATIONAL PARK

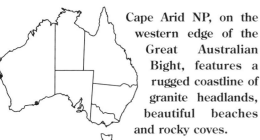

Cape Arid NP, on the western edge of the Great Australian Bight, features a rugged coastline of granite headlands, beautiful beaches and rocky coves.

The shoreline rises gradually through undulating sandplains to the Russell Range. Its granite ridges were once a chain of islands – overhangs and wave-cut platforms just below Mt Ragged's razorback summit mark the ancient sea levels. In spring, a blaze of colour ignites the vast heath plains as tea trees, banksias and myrtle burst into blossom.

The Tagon Coastal Walk, a 15 km return trek, offers magnificent views across the headlands to the Recherche Archipelago. Whales are often seen in late winter to spring. The park's extensive heathlands are easily accessible from the main gravel road across the sandplains to Israelite Bay. Two designated 4WD tracks branch off this road and head north to Mt Ragged and the Eyre Highway. The 3 km climb to the 594 m summit of Mt Ragged offers panoramic views.

Pure white dunes and boulders in Cape Arid NP.

ACCESS

120 km E of Esperance. Conventional vehicle access on unsealed roads to Thomas River and Poison Creek. 4WD tracks elsewhere.

CAMPING

Basic facilities at Thomas River and Seal Creek. No power or water.

WHEN TO VISIT

October to May.

SAFETY

Don't carry plant disease into park. Be aware of tides when taking 4WD on the beach.

ACCESS

About 245 km N of Perth via Brand Hwy and Cervantes Rd. Gravel roads within park are not suitable for caravans or trailers. 4WD tracks to eastern boundary from Brand Hwy and to southern section from Lancelin.

CAMPING

No camping in park. Accommodation and camping facilities at Cervantes about 4 km from park entrance.

WHEN TO VISIT

August through October for walking and wildflowers. Summer is hot and dry.

SAFETY

Take drinking water. Advise ranger of bushwalking plans.

Gould's Monitor.

The limestone pillars of the Pinnacles are seen at their best in early morning or late afternoon light.

A low plateau of ancient and weathered beach dunes forms the backdrop to Nambung's beautiful coastline and intriguing Pinnacles Desert.

These northern sandplains, rich in quartz and lime, roll inland through coastal heath, emerging briefly to swirl among thousands of limestone pillars before flattening out beneath banksia woodlands. Remnants of long-gone tuart forests stand in the park's deeper valleys, while river red gums follow the Nambung River until it disappears amid permanent waterholes into a series of limestone caves.

It's a moody landscape, bleak in the face of wind-driven rain, riotous in its cloak of spring colours, and haunting at dawn and dusk when the Pinnacles cast long, eerie shadows over rippling sand.

An 18 km gravel road runs south from the park entrance to Grey and continues as a rough 4WD coastal track to Lancelin. Two side tracks lead to picnic areas at Kangaroo Point and Hangover Bay, where it's an easy walk to the beach.

A third turnoff heads 6 km east to the Pinnacles Desert. Visitors can drive around a group of these unusual formations on a one-way circuit, or take a 500 m self-guided walk between the pillars, with views across the coast from several lookouts. There are no other walking tracks in the park, so those wanting to see more of Nambung, or do some birdwatching along the river waterholes, should consult the ranger for details on off-track routes.

Coastal exploration is best in calm weather, and there are boat launching facilities at Cervantes for light craft.

This pillar resembles a sculptured figure.

The unique wildflowers of the south-west of Western Australia are at their best in spring and early summer. In the more arid areas of the region, they are at their most spectacular following heavy winter rains.

Nectar-bearing flowers supply food for honeyeaters, lorikeets and other nectar-eating birds; for small marsupials such as Honey-possums; and for bees, butterflies, moths and other insects.

Above left and clockwise: pimelea; yellow flag, orange banksia; scarlet bottlebrush; blue leschenaultia and daisies; mottlecah; fringed lilies; kangaroo paw; scarlet banksia; everlasting daisies.

ACCESS

590 km N of Perth via the Brand and North West Coastal Hwys. Conventional vehicle access on sealed and gravel roads within the park. Scenic flights from Kalbarri township.

CAMPING

No camp sites in park. Bush camping permit required for extended bushwalks. Accommodation and camping facilities at Kalbarri township.

WHEN TO VISIT

August and September, when river levels are high and wildflowers are at a peak.

SAFETY

Carry drinking water and take care when walking near cliff edges.

North of Geraldton, wave-sculpted seacliffs and colourful, river-cut gorge walls provide a dramatic edge to the heathlands of Kalbarri NP.

In August and September, more than 186 000 ha of undulating sandplains produce one of the richest wildflower displays on the west coast. Masses of featherflowers, daisies, mulla mulla, orchids and other ephemeral plants carpet the sandy soils beneath the branches of grevilleas, banksias, melaleucas and mulga.

Spring wildflowers carpet the landscape in Kalbarri NP.

A stone lookout high above a gorge in Kalbarri NP.

The Murchison River runs through Kalbarri NP.

Among the birds attracted by the heathland wildflowers is the Rainbow Bee-eater *(inset opposite)*. However, the park is best-known for its impressive coastal and gorge scenery. The Murchison River has carved long winding gorges through these high coastal plains. For most of the year, the river moves sluggishly through broad reaches and narrow bends at the foot of 150 m cliffs layered with Tumblagooda sandstone and containing marine fossils.

During cyclonic rain, torrents of muddy water sluice through the gorge, spilling into the Indian Ocean from a wide break in the coastal cliffs. South of the river mouth, arched and deeply gullied sandstone cliffs step down to rocky platforms, narrow beaches and offshore pillars.

Two gravel roads, branching off the sealed Ajana-Kalbarri Road, provide access to lookouts and walking tracks along Murchison Gorge. There are great clifftop views over the south-east section of the gorge from Hawks Head and Ross Graham Lookouts.

An easy walk leads to rockpools below Ross Graham. Experienced bushwalkers can continue down the riverbed on a 2 day walk to Z Bend, or carry on from this narrow ravine to an oxbow river bend known as The Loop, to complete a strenuous, 4 day hike covering 38 km. In the course of this walk, several lookouts give magnificent views of the Murchison River as it meanders between dramatic rock walls of warm-coloured sandstone.

Visitors can also drive to lookouts at Z Bend and Nature's Window, where a wonderful rock-framed view of the gorge marks the start of a 6 km walk around the Loop.

West of the gorge, visitors can get an exciting preview of the coastline and Murchison River mouth from the Meanarra Hill summit trail. There are several access points to the park's coastal rock formations and heathlands from a gravel road running south of Kalbarri township.

A 2 hour nature trail at Mushroom Rock offers a self-guided introduction to coastal geology and plant life, while the 8 km seacliff track makes a good half-day walk between Eagle Gorge and Natural Bridge.

Nature's Window, a formation eroded from sandstone.

Nature's Window frames a memorable view of Kalbarri NP.

ACCESS

Conventional vehicle access 850 km N of Perth via Brand and North West Coastal Hwys. 4WD drive only to some areas. Hovercraft from Carnarvon. Aircraft landing strip near Denham. Land and boat tours from Denham. Boat ramps at Denham, Monkey Mia and Francois Peron NP.

CAMPING

Fees charged for bush camping at Steep Point and designated sites in Francois Peron NP. Accommodation and commercial caravan and camping facilities at Nanga, Denham and Monkey Mia.

WHEN TO VISIT

June to October for dolphins, whales and wildflowers.

SAFETY

Carry drinking water when walking. Watch for dangerous marine animals if wading in shallow water.

Stromatolites at Hamelin Pool.

The Shark Bay World Heritage Area covers 23 000 km² of marine and terrestrial habitats on the western extremity of the Australian continent. Its heath and spinifex sandplains drop to a coastline which divides and encloses the shallow waters of Shark Bay with peninsulas and islands, as it snakes south from Carnarvon to the Zuytdorp Cliffs.

It is an internationally important region, where a remarkable array of plants and animals meet within overlapping climate zones. However, it is the fascinating marine life which makes Shark Bay one of Western Australia's "must see" tourist destinations. There are people-friendly dolphins at Monkey Mia, nesting marine turtles, stromatolite-building microbes, a 10 000-strong Dugong population, and a passing parade of Manta Rays, sharks and migrating Humpback Whales.

Above sea-level, the broad, undulating plains behind Zuytdorp Cliffs link the Peron and Edel Land Peninsulas. Here, dry spinifex country adjoins treed heaths of grevillea, melaleuca, banksia and eucalypt.

These semi-arid habitats, and those of Dirk Hartog, Dorre and Bernier Islands, support a variety of animals, including several endangered species, such as the Western Barred Bandicoot and the Banded Hare-Wallaby. Some of the threatened species are being reintroduced to Peron Peninsula, north of Taillefer Isthmus, where fencing has helped to eradicate feral animals and to restore native vegetation.

Saltwater pools such as the one shown above may be nearly landlocked.

The Zuytdorp Cliffs are named after a Dutch ship wrecked on the coastline in 1712.

An aerial view of part of the Peron Peninsula.

THE DOLPHINS OF MONKEY MIA

A small group of Bottlenose Dolphins *(above)* arrive in the shallows at Monkey Mia *(right)* regularly in winter but less frequently in summer. They are fed on an irregular basis so they do not become dependent on human handouts.

Birdwatching is excellent around Shark Bay's grassland, heath, mangrove and seashore habitats. A self-guided, scenic drive takes visitors from the highway into the heart of this World Heritage Area. First stop is the Hamelin Pool stromatolites, structures built by a kind of micro-organism which may have existed for around 3.5 billion years. The drive continues to the shell beach at Lharidon Bight and on to the Eagle Bluff Lookout for views over Freycinet Reach to Edel Land Peninsula and Dirk Hartog Island.

There are two short walks at Denham, and the Lagoon Point trail offers good birdwatching on a 1–2 hour circuit to a shallow claypan at Little Lagoon, returning via the beach.

Aside from the historic Peron homestead near Little Lagoon, it's 4WD only into Francois Peron NP for good fishing, snorkelling and off-track bushwalking.

The unsealed Useless Loop Road cuts across the base of Peron Peninsula to the west coast. About 25 km from the Denham turnoff, the road passes through beautiful spring-flowering heathland.

Visitors need a 4WD to go beyond the magnificent seacliff views at Zuytdorp Point to Steep Point on the tip of Edel Land Peninsula.

ACCESS

285 km S of Port Hedland via the Great Northern Hwy and Wittenoom. Conventional vehicle access on gravel roads from Roebourne and Nanutarra. Roads may be cut after heavy rain.

CAMPING

Fees charged. Basic facilities at Dales, Weano and Joffre Gorge camping areas.

WHEN TO VISIT

May to October. Summer temperatures can exceed 40°C.

SAFETY

Carry drinking water. Advise park staff if undertaking any off-track walks.

The Hamersley Range rises from plains covered with spinifex, termite mounds and eucalypts.

Karijini National Park in the central Pilbara spreads across 628 000 ha of the stark and rugged Hamersley Range. The range is actually a massive, iron-rich plateau stretching for over 400 km between the inland deserts and the Indian Ocean.

An atmosphere of isolation and antiquity pervades its deeply eroded landscapes. Barren escarpments and deep, incredibly narrow gorges cut into the plateau, revealing finely-layered rock formations over 1000 million years old.

The plateau, foothills and valleys appear harsh and dry with a scant covering of spinifex grasses, saltbush and mulga. But summer cyclones and sporadic winter rains bring a flush of short-lived wildflowers, and run-off floods the gorges' normally dry creek beds.

The tiers of banded rock in Weano Gorge.

As the countryside quickly dries, life within the gorges remains comparatively lush. Ghost gums and figs soak moisture from cliff-face fissures. Ferns and mosses grow where moisture trickles over rock ledges, while permanent rockpools and stands of river red gums and paperbarks line the gorge floors.

The park's many habitats support a variety of wildlife, but most species avoid the daytime heat. Reptiles shelter among the rocks, and tall mounds protect grass-eating termites. The Pebble-mound Mouse retreats to tunnels concealed by stones it has heaped up, and other tiny mammals hide under spinifex tussocks.

In springtime, flowers cover the red earth.

Kangaroos, euros and dingos are often seen around waterholes and creeks at dawn and dusk. These small oases are also havens for pythons, frogs and many birds, including pigeons, wrens and finches.

Three groups of Aborigines have been calling Karijini home for over 20 000 years. Today they are involved in management of the park, and offer fascinating insights to the land and its life forms on guided walks.

Visitor facilities and walking tracks are concentrated around the spectacular gorges along the north-eastern escarpments. Visitors can choose from a range of walks and off-track routes classified according to difficulty.

There are several short clifftop walks, with Oxer Lookout providing a hair-raising view down precipitous rock walls to the junction of Hancock, Joffre, Red and Weano Gorges. These four narrow chasms, along with Knox Gorge, are the setting for some of the park's most exciting walks, with the challenge of rock climbs and swims across deep pools.

Further east at Kalamina and Dales Gorges, there are pleasant and very pretty valley floor walks where dense vegetation overhangs crystal clear rockpools. South of the gorges, the 1235 m summit of Mt Bruce offers magnificent views across the plateau.

PLANTS OF THE PILBARA

Above and clockwise: Maidenhair fern growing near water; a ghost gum stands on the spinifex plains; hakea and mulla mulla blossom after rain falls.

At the western end of Dales Gorge, Fortescue Falls feed cool pools of water and lush vegetation.

ACCESS

Coral Bay is 230 km N of Carnarvon via North West Coastal Hwy and Minilya Rd. Exmouth is a further 150 km and the road continues around the cape to Cape Range NP. 4WD coastal track from Yardie Creek to Coral Bay. Boat ramps at Bundegi and Tantabiddi creek. Diving, fishing and sightseeing tours from Coral Bay, Exmouth and Bundegi.

CAMPING

Coastal camping sites include Cape Range NP. Accommodation, caravan and camping facilities at Coral Bay and Exmouth.

WHEN TO VISIT

March—May for Whale Sharks.
July—November for Manta Rays.
August—October for whale-watching.

SAFETY

Wear thick-soled shoes when reef walking. Never dive alone.

Humpback Whales swim past the reefs of Ningaloo MP on their way to their breeding grounds just to the north.

The clear, tropical waters of Ningaloo Marine Park encompass a 260 km barrier of coral reefs. With an offshore span averaging 20 km, the park curves around from Exmouth Gulf and runs south just past the Tropic of Capricorn.

Over 220 species of soft and hard coral and an endless list of algae, fish, molluscs, crustaceans and other marine life forms contribute to the complex interactions that maintain this coral reef ecosystem. It is the larger, itinerant species plying the outer reef edges which attract most attention. The mass spawning of reef corals after the March and April full moons coincides with a seasonal algal bloom attracting plankton-feeding Manta Rays and Whale Sharks. The giant, but harmless, Whale Sharks congregate around Tantabiddi and Mangrove Bay. The plankton drifts also attract schooling fish, which in turn are eaten by big game fish such as marlin.

Manta Rays feed on tiny organisms known as plankton.

From June to November, Humpback Whales can be seen heading to and from their calving ground just north of the park. Green, Hawksbill, Flatback and Loggerhead Turtles start arriving in September, but summer heat and cyclones mean that most visitors miss out on the peak nesting season. By April, the last of the surviving hatchlings has headed north for the winter.

Ningaloo is a diver's paradise. The warm water is free from river sediments and a cast of thousands offers great photo opportunities. Around Mandu Mandu, the reefs are a mere stone's throw from shore and there's plenty of good leeside snorkelling in shallow lagoons. Even the most distant reefs are less than 10 km from the coast, so it's not difficult to reach the outer edges.

Non-divers can see the reef through glass-bottomed boats, enjoy swimming and beachcombing along sandy bays, or take reef walks at low tide.

The Whale Shark is huge, but harmless to humans.

Mt Augustus is reputedly the world's largest rock. To the Wadjari Aborigines, it is a slain ancestral figure known as Burringurrah.

Shrubby mulga, gidgee and myall wattles provide sparse cover over the park's red sandplains and rocky foothills which rise to a 1106 m summit. Good rain brings a flush of ephemeral flowers such as mulla mulla, everlasting daisies and native foxglove.

The northern face of this great monolith drops to the Lyons River valley where flocks of parrots and waterbirds gather at permanent pools shaded with river red gums. Cattle and Edithanna Pools just outside the park are good spots to enjoy birdwatching and a refreshing swim.

A 49 km scenic drive along the park's boundaries offers all-round views of Mt Augustus. Wadjari rock paintings and engraving feature on several short walks on the southern side of the park.

Birdlife around waterholes is a feature of this park.

ACCESS

490 km NE of Carnarvon on unsealed roads via Gascoyne Junction and Mt James. Alternative route via Kennedy Range NP and Lyons Rd. Conventional vehicle access except after heavy rain.

CAMPING

No camping in park. Accommodation, caravan and camping facilities within 5 km of northern boundary.

WHEN TO VISIT

May to September, as summers are extremely hot.

SAFETY

Take drinking water. Advise park staff if climbing to the summit. Avoid tackling unsealed roads in conventional vehicles after heavy rain has fallen.

During the wildflower season, June through to August, is a good time to visit this park.

Getting to the top of the rock is an arduous 12 km return walk from the Beedoboondu carpark and takes at least 6 hours. A shorter walk of 6 km from Oorambu, which leads part way up the northern face, affords impressive views over the Lyons River valley to the Godfrey Range escarpments.

Mt Augustus is definitely off the beaten track, but makes a worthwhile extension to a visit to Kennedy Range NP. Outback adventurers can continue to the Great Northern Highway through the Waldburg and Robinson Ranges in the upper Gascoyne River catchment.

Wildflowers cover the ground after good rains.

DEVONIAN REEF PARKS

ACCESS

Tunnel Creek: 180 km E of Derby or 115 km NW of Fitzroy Crossing. Inaccessible during the wet season.

Windjana Gorge: 145 km E of Derby or 150 km NW of Fitzroy Crossing. Inaccessible during the wet season.

Geikie Gorge: 20 km N of Fitzroy Crossing or 280 km SE of Derby.

CAMPING

Fees charged for good facilities at Windjana Gorge. No camping in Tunnel Creek or Geikie Gorge.

WHEN TO VISIT

May through September. Summer is very hot and may see heavy cyclonic rain.

SAFETY

Be careful when walking on wet rocks. Freshwater Crocodiles are harmless unless provoked.

Tunnel Creek NP.

About 400 million years ago, shallow tropical seas covered much of the Kimberley region in north-western Australia. And microscopic, lime-secreting organisms were fringing its exposed southern highlands with long barrier reefs.

These Devonian-aged reefs now form a series of limestone ranges standing 50–100 m above the surrounding plains and stretching 300 km west from Fitzroy Crossing. Ancient river systems have cut through the ranges, revealing multi-hued cliffs that have been honeycombed, pitted, tunnelled and fissured by groundwater. Summer monsoons now sculpt the Oscar and Napier Ranges as floodwater scours the gorge walls and seeping rainwater slowly dissolves the porous limestone.

During the dry season, when the Lennard and Fitzroy Rivers have contracted to gorge reflecting pools, visitors can explore these fossilised barrier reefs at Windjana Gorge, Tunnel Creek and Geikie Gorge. Access within these small parks is limited, but the scenery is magic and visitors can see relics of a world that existed long before reptiles and mammals roamed the earth.

Sunset illuminates the cliffs of Windjana Gorge.

TUNNEL CREEK NP

Visitors can get an inside look at one of the Kimberley's ancient reefs at Tunnel Creek NP in the Napier Range. Here, a tributary of the Lennard River has created a 750 m long tunnel running from one side of the range to the other. When conditions are favourable, visitors, aided with torches and prepared to wade through cold water, can walk along a tunnel about 15 m wide and 3–12 m high. Halfway through, a roof collapse allows sunlight to infiltrate this subterranean world.

Five species of bats roost among the crevices and small stalactites on the tunnel roof, including fruit-bats, carnivorous Ghost Bats and golden-furred horseshoe-bats.

WINDJANA GORGE NP

The limestone columns of Windjana Gorge flank a 4 km section of the Lennard River where it dissects the Napier Range.

The deeply weathered ridges and walls of this fossilised reef are riddled with caves and tunnels, some containing the most southern examples of rock art depicting Wandjina figures.

These days, aside from raging summer floods, the gorge is a tranquil retreat for waterbirds, parrots, fruit-bats and Freshwater Crocodiles. In autumn, water levels are usually high enough to negotiate the river by canoe and anglers can try their luck with the freshwater fish.

Later in the season, visitors can take a 3.5 km walk into the gorge where tropical vegetation

The Fitzroy River flows through an ancient limestone reef.

The placid waters of Geikie Gorge NP.

fringes sandy river banks and long pools reflect vertical cliffs rising to 90 m. The gorge walls, embedded with fossils, tell much about the development of Australian animals through periods of great climatic change.

GEIKIE GORGE NP

The Fitzroy River breaches a barrier of Devonian reefs near the junction of the Geikie and Oscar Ranges, creating a dramatic gorge.

When 16 m high floodwaters are not leaching and undercutting the deeply etched cliffs, the Fitzroy meanders through long, deep pools and broad sandy reaches. Leichhardt trees, river red gums, pandanus and creeping vines form small forests along riverbanks separated by sheer limestone walls.

The park's wildlife is as varied as its habitats. There are rock-wallabies and Euros among the jagged clifftops and Freshwater Crocodiles, stingrays and barramundi in the river. Waterbirds forage in the shallows, and fruit-bats and parrots feed in the fringing forests. Agile Wallabies, Budgerigars and sand goannas frequent the grassy woodlands surrounding the ranges.

The best way to see Geikie Gorge is on one of the boat tours conducted by park staff or Bunaba Aboriginal guides who have access to restricted areas. After 4:30 p.m., visitors are welcome to explore the gorge in their own boat or canoe, but must not enter the sanctuary areas.

Lennard River Falls.

119

ACCESS

Off Gibb River Road between 180 and 210 km NE of Derby. 4WD recommended. Dry weather access only. Air strips at Silent Grove and Mt Hart homestead.

CAMPING

Basic facilities at Bell Gorge and Silent Grove. Accommodation at Mt Hart.

WHEN TO VISIT

May through September.

SAFETY

Carry adequate supplies of fuel, food and drinking water.

Boab trees on the plains.

About 1750 million years ago, the southern edge of the Kimberley Basin was transformed into a belt of high mountains. Fast-flowing rivers draining the inland cut through the mountains to the ocean. Layers of sandstone were washed away and the King Leopold Ranges were reduced to worn hills and shallow valleys.

The WA government is establishing a 370 000 ha park in the north-western section of the ranges where Bell Creek and the Lennard River drop through terraced falls into narrow gorges. Above the gorges, moderate rainfall supports a scattering of eucalypts, kurrajong and boabs on grass-covered flats and hills which rise to granite outcrops such as Mt Hart.

At present, three tracks lead into this former grazing property. Bell Gorge is the park's best-known feature. Tree-lined Bell Creek follows a valley through the ranges, taking a tumble over layered sandstone slabs into a deep rockpool.

The 50 km drive into the old Mt Hart homestead shouldn't be overlooked, for it offers southside views of the range and its rocky outcrops. Difficult access on a 6 km track makes Lennard Gorge one of the region's least-visited features. It's a steep climb down into this spectacular sandstone feature which flanks the Lennard River for 5 km. A downstream through-walk to Windjana Gorge has the potential to be one of the Kimberley's most challenging bushwalks.

Bell Creek spills over a series of rock ledges before splashing into a deep rockpool.

An Aboriginal painting of a Short-beaked Echidna.

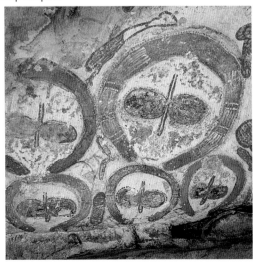

Part of an Aboriginal rock art gallery.

The awe-inspiring domes of Purnululu NP seen from the air.

Purnululu, in the East Kimberley, is one of Australia's oldest landscapes. Over millions of years, erosion has stripped away overlying and surrounding sediments and cut the soft sandstone into a maze of narrow canyons, fractured cliffs and prominent, many-banded domes.

A light aircraft or helicopter ride over the park is the best way to appreciate these remarkable sandstone formations set amidst dry sandplains and shallow river valleys.

Most of the park and the associated nature reserve are closed to the public, but those willing to brave the incredibly rough 55 km trip into the western section can walk among its domes and gorges.

Just past the ranger station, the Spring Gully access road branches north along the floor of the cliff-edged Osmond Valley. Short walking tracks of 1.5–3 km at Echidna Chasm, Mini Palms and Froghole allow visitors to enter narrow chasms hiding permanent waterholes and remnants of tropical rainforest.

On the southern edge of the range, Piccaninny Creek cuts a meandering 12 km gorge between tiers and snaking lines of orange- and grey-banded domes. Visitors taking all or part of the 30 km return gorge walk should advise park staff, as it is easy to get lost among the domes. A 1 km trail near the gorge entrance explores one of the many side canyons as it follows a tributary creek to Cathedral Gorge.

ACCESS

The Spring Gully Rd turnoff is 250 km S of Kununurra or 109 km N of Halls Creek via Great Northern Hwy. 4WD access only. Caravans not permitted. Park closed from 1 January to 31 March. Scenic flights from Kununurra, Halls Creek and from the park's airstrip/helipad.

CAMPING

Fees included with entry pass. Basic facilities with provision for generators at Kurrajong and Walardi. Tour groups only at Bellburn. Bush camping permitted on Piccaninny Gorge walk.

WHEN TO VISIT

May through September.

SAFETY

Carry drinking water when walking. The rock formations are extremely fragile and climbing is not allowed.

Creeks in the gorges may contain water only in summer.

The domes of Purnululu are actually quite fragile formations.

Eastern Grey Kangaroo on roadside.

DRIVE AND SEE

Some wild creatures are easy to spot from a moving vehicle. Drivers should pay attention to the road while passengers scan for wildlife. Look for birds on bare branches, flying across the track or swimming in water beside the road. Watch for birds of prey high in the sky. Look for kangaroos grazing on the plains and road verges from late afternoon onwards (and watch for them hopping across the road as well). See snakes and lizards sunning themselves on the warm road during the day and find frogs and birds such as frogmouths sitting on the track at night.

Sulphur-crested Cockatoo on the alert.

BIRDWATCHING

To make birdwatching even more enjoyable, buy the best binoculars you can afford and a good field guide. Wear comfortable clothing, in layers so it can be shed as the day warms up. Take a hat, sunscreen and insect repellent. Set out as early as possible, as birds are most vocal and visible around dawn. Educate ears as well as eyes, for bush birds often betray their presence by song. Compile locality and lifetime lists, pursue rarities, specialise in one group of birds and/or campaign to save rare birds' habitats.

FINDING WILDLIFE

Visitors to any national park are pleased and excited to see wild creatures in their natural habitat. Sometimes the animals may be comparatively tame, like the kangaroos, possums, kookaburras and goannas which visit camp sites in some parks. At other times, they may be shy and elusive and require dedicated efforts to spot them.

It adds to the enjoyment of a national park visit to know in advance what animals are likely to be discovered there. Readily-available guide books to Australia's birds, mammals, frogs, reptiles and insects should give a general idea of what to look for and where the animals are likely to be seen.

There may also be more specific local guides available.These may be pamphlets, booklets or notice boards in prominent positions within the park or in an information centre.

Some parks offers guided walks on which rangers point out interesting creatures. Many people visit national parks on tours led by birdwatchers or by other experts in the natural history of the area.

WHERE TO SEE SOME FAVOURITES

The Koala is found in eucalypt woodland growing on fertile soil in eastern Qld, NSW, Vic and on Kangaroo Island, SA.

Wedge-tailed Eagles may be seen anywhere they can find small mammals or road kills on which to feed.

The Australian Pelican will be seen anywhere there are large areas of salt or fresh water. It may nest on inland lakes.

Tasmanian Devils roam bushland throughout Tasmania.They are harmless to humans and may visit park camp sites.

The Red Kangaroo is seen on plains and grassy tablelands west of the Great Dividing Range.

The Common Wombat is found in south-eastern Australia. Two rare wombat species live in SW SA and in central Qld.

EUCALYPT WOODLANDS

Australia's eucalypts range from the towering Mountain Ash, Red Tingle and Karri of temperate wet forests right down to the stunted, many-trunked desert mallees. Binoculars are useful when scanning the canopy for birds (honeyeaters and lorikeets flock to eucalypt flowers). During the day, look for dozing Koalas in tree forks and on branches. In spring and summer parrots and other birds nest in tree hollows. Gliders, possums and other mammals take refuge in cavities during daytime, then roam the treetops after dark.

ARIDLANDS

Even in winter it can be extremely hot in Australia's aridlands. Waterholes and billabongs are good places to spot seed-eating birds such as parrots and finches, emus and kangaroos. Most prefer to drink early in the morning or towards evening. Look for reptiles basking in early sunlight and remember that small mammals and lizards lie up during the heat of the day. After heavy rain, look for frogs and waterbirds near temporary water, and watch patches of vegetation for nesting birds such as chats, fairy-wrens and honeyeaters.

RAINFORESTS

In the rainforest, whether it be tropical, sub-tropical or temperate, wildlife can be common but extremely difficult to see. Begin searching at ground level, moving quietly and looking for pademelons, ground-living birds, insects and litter-living frogs. Don't neglect rainforest streams, and look for lizards on tree-trunks, small mammals in tree-ferns. The canopy with its flowers and fruit attracts pigeons, parrots and honeyeaters. At night the rainforest is a magic place and a good torch should reveal possums and nocturnal birds.

SEASHORES

Australia's coastal national parks include some of the continent's most spectacular cliffs, headlands, beaches and mangrove forests. The marine parks which encompass the Great Barrier Reef, Ningaloo, Shark Bay and the Great Australian Bight protect some of the world's most important marine ecosystems. Coastal mudflats and estuaries are havens for migrating shorebirds, while islands and secluded beaches form sanctuaries for breeding and roosting seabirds. Binoculars are essential for spotting seabirds and waders.

A Galah entering its nest in a eucalypt branch.

Eucalypt forest in the Great Dividing Range.

A Red Kangaroo drinks at a billabong.

A waterhole in central Australia's aridland.

The Green Ringtail Possum lives only in tropical rainforest.

Amongst the palms of the Daintree NP.

Crested Terns are found around Australia's seacoasts.

Along Australia's southern coastline.

Aboriginal art should be respected.

A HERITAGE

Many national parks, especially in northern Australia, have galleries of art created by the Aboriginal people. These are of religious and cultural significance and should be treated with respect. Some of the artwork is not intended for particular groups of people to see. If in doubt as to whether photographs should be taken and later shown around, it is best to err on the side of caution and courtesy. In parks administered by the Aborigines, there are usually notices posted regarding their wishes as to access and photography.

Ferns photographed from above with a wide-angle lens.

THE USUAL VIEWS

Photograph from the usual lookouts and viewing platforms in national parks and reserves by all means, but try for some unusual viewpoints as well. Without risking your neck or your equipment, or inconveniencing other people (and respecting any no-go notices), walk a little, climb a little, frame the subject in foliage, use an ultra-wide or extra-long lens if you have one. Get up early for dawn shots, stay out late for afterglow beauty. Pictures are conceived in the mind of the photographer.

PHOTOGRAPHY IN NATIONAL PARKS

To make the best pictures of landscapes, plants and animals, a photographer needs to be very conscious of light.

Time of day, weather and situation all affect light levels. The best photographs are often taken when light levels are quite low, if the camera is held steady so that camera-shake does not blur the image. Marvellous photographs can be taken at dawn and dusk, or in the dim light of the rainforest, or in overcast and stormy weather if a tripod is used, or the camera is rested on a rock or other firm base.

To take great landscape photos it is not necessary to have expensive equipment. An inexpensive 35 mm camera with the best lens in the photographer's price-range should produce very satisfactory shots.

Wildlife is undoubtedly easier to capture on film with the help of a telephoto lens. The rule for using telelenses is the same as for low light conditions – use a tripod to avoid camera-shake. The telelens is also useful for isolating particular elements in landscapes, such as waterfalls or unusual rock formations.

The domes of Purnululu seen over waterworn stones. A wide angle lens was used from an unusually low viewpoint.

A telephoto lens was used to make this dramatic shot of the setting sun and a coastal waterfowl habitat.

TRY THESE TIPS

- Check that your camera (or your video) is working and that you know how to use it before leaving home. Carry plenty of film with you.
- Don't leave camera or film in direct sunlight or in hot places. On the trip, buy film only if it has been kept cool by the retailer. Get film processed as soon as possible after taking it.
- Check camera settings and battery levels each time the camera is used.
- Use a beanbag to cushion a camera on a car windowsill (turn off the motor too).

- If there is strong sunlight and deep shade in one picture, decide which area is most important and focus on it. Then be prepared for other areas to be too bright or too dark.
- If photographing moving animals, track the subject into the frame, take the picture then follow-through smoothly.
- Keep your camera handy when driving.
- If you see a lovely landscape, striking sunset or unexpected wildlife STOP and take the picture. You may never have the opportunity again!

Above: A late afternoon image of Sugarloaf Rock, Leeuwin-Naturaliste NP. The sun is hidden by cloud, creating a subdued but dramatic blue light. *Below:* Boab trees become silhouettes in a photograph exposed for the sunset.

Above: To avoid the picture being too dark, the photographer has taken the exposure on the rock, not on the snow. *Below:* A tripod and a lengthy exposure time result in sharp rainforest and water whose mistiness implies movement.

Kata-Tjuṯa, photographed at dawn so that each clump of spinifex is rounded by shadows and the whole image has a warm, golden tone. The three desert oaks at right balance the great rocks and give depth to the picture. They are echoed in miniature by the faraway trees on the left hand side.

SOME SEASONS AND PLACES

DRIVE TO ARRIVE

Unsealed roads demand full driver attention. Beware loose gravel and soft road verges, bulldust, and dips and humps with impeded view. Find out the depth of water across the road before entering.

BEWARE WILDLIFE

From evening onwards, wildlife poses a traffic hazard. Animals blinded by headlights will freeze. Drivers need to be especially alert for animals grazing on road verges that might veer on to the road in fright.

TIMBER!

Where national park roads coincide with forestry operations, watch out for logging trucks. One-lane tracks through mountain forest, especially with creek crossings, are a test of drivers' skills.

CARE ABOUT KIDS

Take advantage of every park facility to let kids stretch their legs, refresh themselves and use the toilets. Be aware that wild creatures should be admired but neither patted nor fed.

SUMMER

Summer in the Top End is hot and humid, but exciting.

Tasmania's national parks are visitor-friendly in summer.

AUTUMN

After summer rain, North Queensland's rainforest flourishes.

Try the Great Barrier Reef after summer's cyclones end.

WINTER

Australia's alpine national parks are winter wonderlands.

Glorious winter weather adds to the Red Centre's charm.

SPRING

Wildflowers turn south-west WA into a garden in spring.

Spring is a great time to see Victoria and South Australia.

USEFUL ADDRESSES AND TELEPHONE NUMBERS

TOURIST OFFICES

Australian Capital Territory Tourism Commission, Level 13 Saphouse, Bunda & Akuna Sts, Canberra 2601 (tel. 1800 026166 toll-free or 02 62205 0666) www.canberratourism.com.au

Tourism New South Wales, 11–13 York St, Sydney 2000 (tel. 13 20 77) www.tourism.nsw.gov.au

Northern Territory Tourist Commission, 43 Mitchell St, Darwin 0801 (tel. 13 30 68) www.nttc.com.au

Tourism Queensland, 30 Makerston St, Brisbane 4000 (tel. 07 3535 3535) www.tq.com.au

Tourism Tropical North Queensland, 51 The Esplanade, Cairns 4870 (tel. 07 4031 7676)

South Australian Tourism Commission, 18 King William St, Adelaide 5000 (tel. 08 8463 4500 or 1300 655 276 toll-free) www.southaustralia.com

Tourism Tasmania, 22 Elizabeth St, Hobart 7000 (tel. 03 6230 8235) www.tourismtasmania.com.au

Tourism Victoria, 55 Collins St, Melbourne 3000 (tel. 13 28 42) www.visitvictoria.com

Western Australian Visitor Centre, 469 Wellington St, Perth, WA 6000 (tel. 1 300 361 351) www.westernaustralia.net

ACCOMMODATION

Bed & Breakfast Australia (head office), 29 Burlington Rd Homebush NSW 2071 (tel. 02 9763 5833) www.bedandbreakfast.com.au

Youth Hostels of Australia, (head office), 422 Kent St, Sydney NSW 2000 (tel. 02 9261 1111) www.yha.com.au

Backpacking Around, PO Box 1058 Cloverdale WA 6985 (tel. 08 9277 6193) www.backpackingaround.com.au

* All addresses and telephone numbers on this page are subject to change. If problems arise, check address and contact number in current telephone directory for city.

MOTORING CLUBS

There is an automobile association in each Australian State and Territory. When driving anywhere in Australia, it is useful to be a member of one of these organisations, for each organisation is affiliated with the Australian Automobile Association, and when interstate a member of any organisation can request services from the local body.

Some of the popular facilities offered are emergency breakdown assistance, vehicle towing and vehicle inspection services. Other services include tuition in safe and defensive driving, motor vehicle insurance cover and advice as to approved repairers. Extremely useful membership benefits include advice about motoring holidays and tours. This can consist of maps, guides, reports on road conditions, travel bookings, concessional rates for accommodation and car hire, and other services.

Club, or social, membership allows the use of club facilities and accommodation.

**Phone 13 11 11
for emergency roadside assistance
throughout Australia**

AAA Australian Automobile Association, 216 Northbourne Ave, Braddon Canberra, ACT 2612 (tel. 02 6247 7311)

AANT Automobile Association of the Northern Territory, 79-81 Smith St, Darwin, NT 0800 (tel. 08 8981 3837)

NRMA National Roads & Motorists Association, 388 George St, Sydney, NSW 2000 (tel. 02 9292 9222)

RAA Royal Automobile Association of South Australia, 41 Hindmarsh St, Adelaide, SA 5000 (tel. 08 8202 4600)

RACQ Royal Automobile Club of Queensland, 300 St Pauls Tce, Fortitude Valley, Qld 4006 (tel. 07 3361 2406)

RACT Royal Automobile Club of Tasmania, cr Patrick & Murray Sts, Hobart, Tas 7001 (tel. 03 6232 6300)

RACV Royal Automobile Club of Victoria, 550 Princes Hwy, Noble Park, Vic 3174 (tel. 13 19 55)

RACWA Royal Automobile Club of Western Australia, 228 Adelaide Tce, Perth, WA 6000 (tel. 08 9421 4400)

NATIONAL PARKS

Environment Australia, (head office) John Gorton Building King Edward Terrace, Canberra, ACT 2601 (tel. 02 6274 1111)

Australian Capital Territory Parks & Conservation Service, PO Box 1119, Tuggeranong, ACT 2901 (tel. 02 6205 1233)

New South Wales National Parks & Wildlife Service, 43 Bridge St, Hurstville, NSW 2220 (tel. 02 9585 6444)

Parks & Wildlife Commission of the Northern Territory, 38 Cavanagh St Darwin, NT 0830 (tel. 08 8999 5511)

Parks Australia North, GPO Box 1260, Darwin NT 0801 (tel 08 8946 4300) (Uluru information tel. 08 8956 2299; Kakadu information tel. 08 8938 1100)

Environmental Protection Agency, 160 Ann St, Brisbane 4000 (tel. 07 3227 7111)

Great Barrier Reef Marine Park Authority, 2-68 Flinders St, Townsville, Qld 4810 (tel. 07 4750 0700)

Department of Environment & Heritage, 77 Grenfell St, Adelaide, SA 5000 (tel. 08 8204 1910)

Department of Primary Industries, Water & Environment , GPO Box 44, Hobart, Tas 7001 (tel. 03 6233 8011)

Parks Victoria, 535 Bourke St, Melbourne, Vic 3000 (tel. 13 19 63)

Department of Conservation & Land Management (CALM), 50 Hayman Rd, Como, WA 6152 (tel. 09 9334 0333)

EMERGENCY!
IN ALL STATES
FOR POLICE,
AMBULANCE AND
FIRE-BRIGADE
DIAL 000

Publisher's note: While every effort has been made to ensure that the information in the book is accurate at the time of going to press, things change and the Publisher cannot accept responsibility for errors or omissions. Steve Parish Publishing welcomes information and suggestions for corrections and improvements from readers.

INDEX